VoIP EMERGENCY CALLING

VoIP EMERGENCY CALLING

FOUNDATIONS AND PRACTICE

Karl Heinz Wolf
nic.at, Austria

Richard Barnes
BBN Technologies, USA

WILEY
A John Wiley and Sons, Ltd., Publication

This edition first published 2011

Registered office
John Wiley & Sons Ltd, The Atrium, Southern Gate, Chichester, West Sussex, PO19 8SQ, United Kingdom

For details of our global editorial offices, for customer services and for information about how to apply for permission to reuse the copyright material in this book please see our website at www.wiley.com.

Library of Congress Cataloging-in-Publication Data

Barnes, Richard, 1982-
VoIP emergency calling : foundations and practice / Richard Barnes, Karl Heinz Wolf.
p. cm.
Includes bibliographical references and index.
ISBN 978-0-470-66594-7 (cloth)
1. Internet telephony. 2. Telephone–Emergency reporting systems. I. Wolf, Karl Heinz, 1982- II. Title.
TK5105.8865.B36 2011
384.6′4—dc22

2010033579

A catalogue record for this book is available from the British Library.

Print ISBN: 978-0-470-66594-7 (H/B)
ePDF ISBN: 978-0-470-97696-8
oBook ISBN: 978-0-470-97697-5
ePub ISBN: 978-0-470-97694-4

Typeset in 11/13pt Times Roman by Laserwords Private Limited, Chennai, India

Contents

Foreword

In an emergency, it has become natural to reach for the nearest telephone in order to obtain assistance from Law Enforcement, the Fire Department, or the emergency services (EMS). From any common fixed or mobile telephone, you dial an emergency number, and the telecommunications provider routes the emergency call in such a way that it will be delivered to the control center for the proper emergency service. The caller's number is sent along with the call and the calling location is made available.

With Internet telephony (Voice over IP, or VoIP), the situation is a little more complicated. It is often possible for a caller to dial into the VoIP system and make or receive calls no matter where he is, as long as he has a suitable Internet connection. Making the clear correspondence between a calling number and a geographical location is more difficult (or may not happen at all). For VoIP service operators, the extreme case means failure to deliver an emergency call, since the caller's whereabouts are unknown.

In the United States, for example, it is required that the caller's "registered location" be part of the emergency calling process. This means that the caller has to pre-identify the location of their VoIP interface connection. However, if the caller is nomadic, moving periodically to another service location, and they do not update the emergency service record to reflect their new location, their subsequent emergency call may well be routed to the wrong place. The challenge for the future, then, is how to accomplish automatic location update, based on where the caller really is when they make that critical emergency contact. Looking forward to the day when VoIP emergency calls are supported with the technology to supply current location involves development and convergence of technical and operational standards.

In my role as Technical Issues Director for the National Emergency Number Association (NENA), I welcome the authors' initiative in presenting this publication to give interested readers comprehensive theoretical and practical guidance on emergency calling over the Internet.

Roger Hixson
Technical Issues Director – NENA

Useful Links

Throughout this book there are many references to documents of the Internet Engineering Task Force (IETF). In the jargon of the IETF, RFCs are permanent, archival documents that reflect the consensus of the IETF, while Internet-drafts are working documents (which often eventually become RFCs); all RFCs start as Internet-drafts. All of these documents can be quickly accessed by name using the IETF document retrieval service. The document named "document-name" can be accessed at the URI `http://tools.ietf.org/html/[document-name]`. For example:

- draft-ietf-ecrit-framework can be found at `http://tools.ietf.org/html/draft-ietf-ecrit-framework`
- RFC 5582 can be found at `http://tools.ietf.org/html/rfc5582`

In cases where the document name is unclear, you can also search for author names or titles on the general IETF document site at `http://tools.ietf.org/html/`.

Currently, ECRIT and GEOPRIV are still refining these standards and developing new documents to address a few remaining use cases. There are thus more documents being developed than are referenced specifically in this book, so you can also refer to the working group status pages:

- GEOPRIV: `http://tools.ietf.org/wg/geopriv/`
- ECRIT: `http://tools.ietf.org/wg/ecrit/`
- SIPCORE: `http://tools.ietf.org/wg/sipcore/`

These pages will give you the latest information on the state of play regarding the Internet standards being developed by these groups.

List of Abbreviations

Topics covered in this book involve several different technical areas, standards organizations, and regions of the world. This broad array of topics brings with it a wide array of jargon and acronyms. This list is provided as a reference point for the abbreviations in the main text.

3GPP	Third-Generation Partnership Project
ADSL	Asymmetric Digital Subscriber Line
AJAX	Asynchronous Javascript And XML
AK-TK	Arbeitskreis für technische Koordination (Technical Coordination working group, Austria)
ALI	Automatic Location Identification
ANI	Automatic Number Identification
ANSI	American National Standards Institute
ATOCA	Authority-to-Citizen Alerting (IETF working group)
BAKOM	Bundesamt für Kommunikation (Federal Office of Communications, Switzerland)
BCP	Best Current Practices (IETF document type)
BEREC	Body of European Regulators for Electronic Communications
CDP	Cisco Discovery Protocol
CGALIES	Co-ordination Group on Access to Location Information by Emergency Services
CID	Content-ID
CPE	Customer Premise Equipment
CSCF	Call Session Control Function
DHCP	Dynamic Host Configuration Protocol
DHCPv6	Dynamic Host Configuration Protocol for IPv6
DNS	Domain Name System
DNSSEC	Domain Name System Security Extensions
DOP	Dilution of Precision
DSL	Digital Subscriber Line
ECRIT	Emergency Context Resolution with Internet Technologies (IETF working group)

EENA	European Emergency Number Association
EGEA	Expert Group on Emergency Access
ELIN	Emergency Location Identification Number
EMTEL	Emergency Telecommunications (ETSI working group)
EPSG	European Petroleum Survey Group
ESGW	Emergency Services Gateway
ESInet	Emergency Services IP network (NENA)
ESW	Emergency Services Workshop
ETSI	European Telecommunications Standards Institute
EU	European Union
FCC	Federal Communications Commission (US)
FG	Forest Guide
FMK	Forum Mobilkommunikation (Mobile Communications Forum, Austria)
FTTH	Fiber to the Home
GEOPRIV	Geolocation and Privacy (IETF working group)
GML	Geography Markup Language
GMT	Greenwich Mean Time
GPS	Global Positioning System
HELD	HTTP Enabled Location Delivery
HTTP	Hypertext Transfer Protocol
HTTPS	HTTP Secure
IANA	Internet Assigned Numbers Authority
ICE	Industry Collaboration Event (NENA)
IEEE	Institute for Electrical and Electronics Engineers
IETF	Internet Engineering Task Force
IMS	IP Multimedia Subsystem
IMSI	International Mobile Subscriber Identity
IP	Internet Protocol
IPv4	Internet Protocol version 4
IPv6	Internet Protocol version 6
ISC	Internet Systems Consortium
ISO	International Organization for Standardization
ISP	Internet Service Provider
ITU-T	International Telecommunication Union – Telecommunication Standardization Sector
KEM-V	Kommunikationsparameter-, Entgelt- und Mehrwertdiensteverordnung (Communications Parameters, Fees and Value-Added Services Ordinance, Austria)
LCP	Location Configuration Protocol

LIS	Location Information Server
LLDP	Link-Layer Discovery Protocol
LLDP-MED	Link-Layer Discovery Protocol – Media Endpoint Discovery
LoST	Location-to-Service Translation Protocol
LRF	Location Retrieval Function
MIC	Ministry for Internal Affairs and Communications (Japan)
MIME	Multipurpose Internet Mail Extensions
MLTS	Multi-Line Telephone System
MSAG	Master Street Address Guide
NAPTR	Naming Authority Pointer
NAT	Network Address Translation
NENA	National Emergency Number Association (US)
NG112	Next Generation 112
NG9-1-1	Next Generation 9-1-1
NICC	Network Interoperability Consultative Committee (UK)
OASIS	Organization for the Advancement of Structured Information Standards
OMA	Open Mobile Alliance
PBX	Private Branch Exchange
PEACE	IP-Based Emergency Application and Services for Next-Generation Networks
PIDF-LO	Presence Information Data Format – Location Object
PSAP	Public Safety Answering Point
PSTN	Public Switched Telephone Network
RFC	Request For Comments (IETF document type)
RFID	Radio-Frequency Identification
RTP	Real-Time Protocol
RTR	Rundfunk und Telekom Regulierungs-GmbH (Regulatory Authority for Broadcasting and Telecommunications, Austria)
SBC	Session Border Controller
SDP	Session Description Protocol
SIM	Subscriber Identity Module
SIP	Session Initiation Protocol
SIPCORE	Session Initiation Protocol Core (IETF working group)
SIPS	Session Initiation Protocol Secure
SMS	Short Message Service
SNMP	Simple Network Management Protocol
SQL	Structured Query Language

SRS	Spatial Reference System
TBD	To Be Determined
TCP	Transmission Control Protocol
TIA	Telecommunications Industry Association
TKG	Telekommunikationgesetz (Telecommunications Act, Austria)
TLS	Transport Layer Security
UDP	User Datagram Protocol
URI	Uniform Resource Identifier
URL	Uniform Resource Locator
URN	Uniform Resource Name
UTC	Coordinated Universal Time
UTF-8	Unicode Transformation Format (8-bit)
VDSL	Very-high-bitrate Digital Subscriber Line
VPC	VoIP Positioning Center
VPN	Virtual Private Network
VoIP	Voice over IP
VSP	VoIP Service Provider
W3C	World Wide Web Consortium
WGS84	World Geodetic System, 1984
WLAN	Wireless Local Area Network
XCAP	XML Configuration Access Protocol
XML	Extensible Markup Language
XMPP	Extensible Messaging and Presence Protocol
XPCOM	Cross Platform Component Object Model

1

Introduction

Emergency calling is a critical function of current telephone networks. Emergency calls are placed in order to prevent danger to life and limb, as well as to property and the environment. In an emergency, it's essential that appropriate responders arrive quickly, so an emergency call must be routed directly to the responsible emergency call center. The police in Vienna, Austria alone receive 3,000 to 4,000 calls each day – from fixed-line telephones, mobile phones, and Internet phones.

This "VoIP Emergency Calling" book is focused on emergency calls that are placed using Voice over IP (VoIP) (and thus carried over the Internet), since these calls often receive inadequate service. We will describe the standardization activities of the Internet Engineering Task Force (IETF) devoted to bettering this situation for VoIP emergency calls, and then delve into them in a more practical terms. This book is intended for several audiences interested in emergency services: operators of emergency call centers, VoIP services, and IP networks; vendors of VoIP telephones and software; representatives of regulatory bodies. Those who are focused more on operational considerations will be able to explore several scenarios with the help of the Practice chapter.

It should be noted that work on VoIP emergency calling architectures is still ongoing, and the relevant standards are not final. This could mean that the eventual deployments (according to the final standards) could differ slightly from the discussion in this version of the book. However, the basic standards are essentially complete, so the technical narrative and exercises in this book should provide a good foundation, even if the details of the ultimate standards differ.

In the rest of this introductory chapter, we'll first discuss how VoIP emergency calls differ from those placed over traditional fixed and mobile telephone networks. Later we present an overview of the remaining chapters.

VoIP Emergency Calling: Foundations and Practice Karl Heinz Wolf and Richard Barnes

1.1 Calling over the Internet

We talk about "Internet Telephony" (usually written as VoIP) if a telephone conversation is carried over the Internet. The voice content of these calls is forwarded through the network in small packets, using the wide variety of protocols and platforms currently deployed. Using VoIP, both the caller and the callee can be reachable directly over the Internet, or they can be connected to traditional telephone networks with the help of gateways. VoIP operators typically set up gateways that are available for their customers to use.

For users, all this new technology is often invisible, since an "Internet telephone" can be operated in the same way as a normal telephone, and can even look the same. Indeed, the use of VoIP adapters enables traditional fixed-line phones to make VoIP calls. There are also mobile devices for VoIP – there are many VoIP applications for major smart phone operating systems (in particular, Apple iOS and Android). Some devices, such as the Nokia N810 or the Apple iPod Touch, can *only* make calls using VoIP. It is important to note that a caller doesn't necessarily need to use a VoIP operator in order to make a VoIP call. Calls over the Internet can be set up directly between any number of interested parties.

Internet access is already available almost everywhere in the world, and so VoIP calling is also possible almost everywhere – even using a single "home" VoIP operator. This means that a VoIP user is always reachable with the same telephone number (or e.g., a SIP URI, a VoIP equivalent) – no matter where he is located at the time. This mobility, in the truest sense of the word, respects no borders.

The number of callers using VoIP is continually increasing, and current fixed telephone access networks (the PSTN) will eventually be replaced with VoIP. Indeed, in many areas, local PSTN service is already being replaced with VoIP, for example, as another service running on a Fiber-to-the-Home (FTTH) network. VoIP operators assure their customers that VoIP works exactly like a traditional telephone line, and offer several additional services. Unfortunately, there are problems with emergency calling all too often. The next section will explain the reasons why.

1.2 VoIP Emergency Calling Problem Statement

In order to understand the challenges involved in VoIP emergency calls, it helps to clarify two fundamental requirements for an emergency call:

- The emergency call must be routed to the emergency call center responsible for the caller's actual location.
- The emergency call center needs the location of the caller in order to dispatch first responders.

Emergency calls must therefore be routed – with the caller's location attached – to the responsible emergency call center. Call centers are only responsible for prescribed geographical area. If an emergency call is routed to the "incorrect" emergency call center (i.e., one that isn't responsible for the caller's location), the call must first be forwarded to the responsible call center – while valuable time passes. In addition, there is a danger that the caller will become panicked if he is informed that his call has been misrouted. Only the responsible emergency call center can send the assistance that the situation calls for. For this, the call center requires the most accurate possible location for the caller. In most emergency situations, emergency responders must be sent to the caller's location, so it is essential for emergency response planning that location information is sent all the way to the emergency call center.

Location information is thus necessary for two steps: first, to determine the responsible call center and, second, to determine emergency response planning. In any case, the requirement for location information conflicts with the boundless mobility of VoIP. Location determining is thus the central problem for VoIP emergency calls. Traditional fixed telephones are, of course, not mobile (outside the confined distance a cordless phone can travel). By contrast, one can easily carry a VoIP telephone and use it in another place, as long as Internet access is available, so VoIP telephones are fundamentally mobile.

Why is determining the location of VoIP callers so difficult? The following points are crucial:

- VoIP supports end-user mobility.
- In VoIP systems, there is no correspondence between a telephone number and the caller's location (in many cases no phone number will be assigned).
- On the Internet, transport and voice service are distinct (a caller's network and voice service providers can be different entities).

On the Internet, the separation of transport and voice service is common – in contrast to current telephone networks. A network operator is only responsible for providing Internet access and transporting data (e.g., voice data). Many VoIP operators offer VoIP service that can be reached over any access network. No agreement between the two operators is necessary for someone to use a particular VoIP service over a particular access network.

A caller can find himself in an arbitrary network and use a different VoIP operator. In this situation, the network operator generally has no idea which data are being carried through his network, and thus has no way of knowing that a user would like to make an emergency call. The VoIP operator possibly can recognize an emergency call, however, he usually does not know in which network (or where in the network) the user is currently located. On the other hand, in many situations, the operator whose network a user is connected to is capable of locating the client in the network.

The Internet Engineering Task Force (IETF) is the standards body for the Internet. The IETF developed the emergency calling solutions that are laid out in this book in order to resolve this dilemma between VoIP mobility and the separation of transport and voice service. The problems described above aren't the only ones; there are still other things to consider:

- *Gateways:* Currently, all emergency call centers are only accessible over the traditional telephone system, and not over the Internet (with the exception of a few forward-looking ones). This means that VoIP callers currently need a gateway from the Internet to the PSTN in order to make an emergency call. Most VoIP operators have already deployed gateways; however, a subscription to that VoIP operator is almost always necessary to be able to use their gateway. In any case, it isn't strictly necessary for a VoIP caller to use a VoIP operator.
- *Telephone numbers:* VoIP services don't necessarily require telephone numbers. In any case, emergency call centers can only place calls back exclusively to telephone numbers. It is impossible right now for an emergency call center to place a return call to a VoIP caller without a telephone number.
- *Location conveyance:* Even when location information is available, it cannot be relayed automatically to an emergency call center, since this function isn't provided by the traditional telephone system.

These constraints arise from the fact that most emergency call centers are only reachable over traditional telephone networks and not over the Internet. Modernizing emergency call centers would resolve not only these problems, but also other deficiencies of the current system.

In summary, it's important to keep in mind that location information is absolutely essential for an emergency call (without it, there's no way to deliver the call to the responsible emergency call center) and location information is difficult to determine (due to the mobility of end users and the division between transport and the VoIP service).

1.3 Emergency Communication

An emergency call is frequently the action that initiates an emergency response, but it is seldom the only form of communication that takes place as part of that response. In addition to the emergency call, dispatchers, responders, and other entities need to communicate with each other, and sometimes need to send alerts out to other individuals. In general, there are four types of communications that happen during an emergency:

1. From the individual to the authority.
2. From the authority to the individual.
3. Among authorities.
4. Among individuals.

Emergency calls are placed by an individual to avert threat to life or physical condition and to property by calling an authority, for example, the police. Often during emergencies, and especially during large-scale disasters, authorities have to communicate among themselves, for example, to share information and to coordinate rescue. Sometimes it is also necessary to alert and warn individuals of dangers. Early warning is therefore directed from an authority to individuals. So for the first three forms of emergency communication there is always an authority involved. However, one also has to consider that people affected by a disaster want to communicate with their relatives and friends, for example as described in Conti (2008). Consequently, the fourth type of emergency communication is completely among individuals. All these kinds of communication have to be considered when preparing a communication network for use in emergency situations.

All these kinds of communication can be carried out over the Internet as well (or at least using Internet technologies), which raises the question of how to ensure that devices that implement these functions can interoperate. The need for interoperability is greatest for "individual-to-authority" and "authority-to-individual" cases, because there are a great number of devices involved made by many vendors, with no central coordination. (In the "authority-to-authority" case, there is commonly a single government to coordinate systems.) The "individual-to-individual" case is basically the standard use of Internet applications, which are already designed to be interoperable; some consideration of the robustness of these applications might be warranted, however.

Because the need for interoperability has been greatest in the "individual-to-authority" and "authority-individual" cases, most standards development work has focused on these cases. Work on authority-to-individual communications is just beginning in the IETF, within the ATOCA working group, building on prior work done in the 3GPP, ITU-T, and elsewhere. Citizen-to-authority communications (i.e., emergency calling) received attention first because it is so central to current telephone networks, and because of the urgency created by the rapid transition to VoIP. Emergency calling is the subject of the IETF ECRIT working group, and the focus of the remainder of this book.

1.4 Overview of this Book

This book is intended for anyone interested in the theme of VoIP emergency calling, especially operators of networks and VoIP services, vendors of VoIP hardware and software, and emergency call centers.

In the next part of this book, we will give an overview of the actual state of emergency calling, and the current regulatory situation. Then we'll explore the new emergency calling architecture for VoIP, and round out the book with a section on practical considerations.

Chapter 2 gives a basic overview of emergency calling, the requirements of the emergency calling system and in particular stresses the importance of location information.

Chapter 3 is concerned with the architecture that the IETF has developed VoIP-based emergency calls. Subsections explain the different concepts and parts of the architecture that are necessary for an emergency call.

Chapter 4 is dedicated to the most important information for an emergency call, the caller's location. Location determination and location configuration, with the help of different protocols, are the main content of this chapter.

Chapter 5 overviews the regulatory situation and implementation of the current emergency calling system and regulatory framework in Austria, the European Union, the US, and Japan.

Chapter 6 is the practical chapter of the book. In addition to an introduction to already existing tools for VoIP emergency calls, there are concrete practical guidelines. These examples illustrate the processes and protocols described in previous chapters.

Chapter 7 looks at the security features of the IETF emergency calling architecture, from the perspectives of all the stakeholders involved.

Chapter 8 takes a critical look at the IETF emergency calling architecture and points out some obstacles to implementation.

Chapter 9 concludes the book with summary and a short forecast.

The book is also supported by a companion website available at `http://www.voip-sos.net/`. This webpage contains news related to VoIP emergency calling and the ongoing IETF work. Furthermore, downloads and links to software interesting for VoIP-based emergency calling are provided. So check out this webpage to stay updated.

References

Conti JP (2008) Here, there and everywhere. *Engineering Technology* **3**(14), 72–75.

2

Emergency Calling

Emergency calls are a critical function of the current telephone system; they are no longer an optional feature. In order to get help in an emergency, people intuitively reach for the nearest telephone and dial a well-known emergency number.

Since the rest of this book is devoted to describing the emergency calling system for VoIP, this chapter explains the current situation for emergency calls as well as considerations for Public Safety Answering Points (PSAPs) and the emergency calling system as a whole.

2.1 Overview

In most countries, emergency calls are initiated by dialing a short, memorable string of numbers. There are countries with only one emergency number (as with 911 in the US) and other countries with several numbers (such as Austria, which uses 122, 133, 144, and others). Depending on the country, several different emergency services can be available. In addition to the best-known services – fire, police, and ambulance – in some areas there are emergency services for coastguard emergencies, forest fires, natural gas leaks, and in some cases special police departments. In Austria, for example, there are a series of emergency services available beyond the standard police, fire, and medical services (see Section 5.2 for details).

Emergency calls are connected to Public Safety Answering Points (PSAPs) – these are the "call centers" for emergency calls. There, a call taker will answer the call, and usually ask immediately what has happened and where. Precise answers to these questions make the call taker's job easier, since the emergency must be categorized and the relevant emergency responders dispatched to the location of the emergency. Unfortunately, the answers that callers give to these questions often do not provide enough information for the call taker – the

VoIP Emergency Calling: Foundations and Practice Karl Heinz Wolf and Richard Barnes

question of the caller's location is especially problematic. For this reason, systems that deliver location information to the call taker automatically can be an essential improvement to an emergency calling system. The significance of location information to emergency calls is discussed in Section 2.3 below.

PSAPs can be organized in several different ways, depending on their size, and on how they are governed and regulated. In small PSAPs, the same group of people might handle incoming calls as well as requesting help from first responders and coordinating the response. In larger PSAPs, these responsibilities are usually assigned to different groups: "call takers" answer incoming emergency calls, and "dispatchers" oversee the coordination of first responders. PSAPs can also be organized so that certain emergency calls are forwarded to other, more specialized PSAPs.

Most PSAPs today are reachable only over the traditional telephone system (the PSTN), not over the Internet. A small but growing fraction of PSAPs have access to the Internet so that responders can look up additional information, but they cannot receive emergency calls from the Internet. Some specific requirements that PSAPs have for the emergency calling system are discussed in the following section.

2.2 Infrastructure Requirements

This section summarizes some important requirements for an infrastructure to support emergency calling, especially for PSAPs. These requirements apply in the traditional telephone networks as well as to possible future expansions and improvements to the emergency calling system through Internet technologies.

This summary we present here brings together the critical points from the US National Emergency Number Association (NENA), the IETF, as well as some common regulations. Further requirements can be found in the original documents produced by these groups (*NENA IP-Capable PSAP Minimum Operational Requirements Standard (58-001)*, 2007) and Rosen and Polk (2010) for NENA and the IETF; further regulatory considerations are discussed in Chapter 5).

The most commonly cited requirements for emergency calling in the traditional telephone network are the following:

- Ensuring the reachability of the local emergency calling numbers.
- Connecting callers to the PSAP that is responsible for the caller's current location.

- Providing access to emergency numbers free of charge (including support for callers who have no service).
- Providing PSAPs with information on the identity of the caller (today, via the caller's phone number).
- Allowing PSAPs to call back callers if the connection is dropped or the PSAP requires further information.
- Enabling PSAPs to request subscriber information and location information from networks.
- Giving emergency calls priority in the network.
- Allowing emergency calls even when other calls are not allowed from a given line.

These requirements are naturally also valid for VoIP-based emergency calls as well, although they may implemented in different ways. (See Chapter 3 for an overview of the VoIP emergency calling system, and Chapter 4 for information on location information in particular.) There are also a few inherent properties of the current telephone system that become additional requirements for VoIP:

- Emergency calling must be possible without additional configuration by the user.
- Emergency calls must work without a VoIP provider (for VoIP, a voice provider is not strictly necessary).
- The system must prevent callers from accidentally disconnecting.
- "Silence suppression" features must be deactivated (since background noises can be informative for emergency response).
- Features such as voicemail, call forwarding, call hold, and "do not disturb" must be deactivated during an emergency.

There are also some ways in which Internet technologies can enable improvements to emergency calling systems:

- Forwarding of calls to other PSAPs with all associated data.
- Support for additional kinds of media (such as video and real-time text).
- Special handling for callers in a designated geographical area (enabled by direct delivery of location information).
- Improved conferencing functions.
- Improved collaboration between PSAP workers using Internet technologies.
- Availability of home emergency dial string in addition to local emergency numbers.

The introduction of new types of media, direct delivery of location information, and other Internet technologies represents a significant improvement to the emergency calling system. Naturally, PSAPs will need to be upgraded in order to take advantage of these features. The point is, though, that there are emerging requirements for PSAPs to be reachable over the Internet, and simply supporting standard VoIP technologies will in itself allow PSAPs to support many common media types (e.g., audio, video, and text). Although one speaks of emergency *calls*, these communications are not limited to voice calls. As communication patterns continue to evolve to support additional media, IP-enabled PSAPs will be able to provide significantly improved services, especially for people with disabilities. For example, for speech- or hearing-impaired people, exchanging instant messages with a PSAP would be a great improvement over many current technologies (such as sending faxed messages or SMS), and would likely make the call taker's job easier as well. With the help of video calls and conferencing, sign language can be conveyed and an interpreter conferenced in. Video from the caller's location can help a call taker form a picture of the situation in progress. A video call can also be used to instruct the caller in how to apply first aid.

2.3 The Role of Location Information

Location information is not just a nice-to-have feature for VoIP emergency calls that slightly eases the work at the PSAP. Location information is the key element that enables emergency calls in the first place – not only for VoIP-based emergency calling, but also in the traditional telephony system.

Since call takers at PSAPs typically ask the caller for his current location, one wonders why location information needs to be provided automatically. If the caller can describe his location well enough, perhaps automatic location is not so important. However, in many cases, callers cannot provide sufficient information to call takers, for any of a number of reasons. One could also note that only the enhanced emergency calling system (e.g., E9-1-1) demands that explicit location information be displayed to the call taker (*NENA IP-Capable PSAP Minimum Operational Requirements Standard (58-001)*, 2007). However, the situation is more complex. Location information is actually necessary to detect an emergency call, to route the call, and to send first responders to the site of the emergency.

Table 2.1 A selection of emergency numbers from around the world. Note the re-use of numbers for different services, and the variety of services offered

Number	Service	Country
911	All services	United States
112	All services	European Union
999	All services	United Kingdom
999	Private ambulance	China
122	Fire brigade	Austria
122	Police	Bosnia
111	All services	New Zealand
111	Fire brigade	Sri Lanka
17	All services	France
171	All services	Venezuela
1548	Counter-terrorism	Algeria
113	Report a spy	South Korea
113	Fire brigade	Syria
113	State police	Italy
1414	Helicopter rescue	Switzerland
165	Anti-kidnapping	Colombia
10177	Ambulance	South Africa

The basic PSTN emergency calling system does not convey caller location in call signaling, but nevertheless, the well-known local emergency dial strings (like 9-1-1 for the US or 1-1-2 for most parts of Europe) have to be available and the call has to be routed to the PSAP serving the location of a caller. Emergency dial strings themselves are location dependent and assigned by local authorities (see Table 2.1 for a sample of numbers around the world). Hence, location information is necessary to figure out the local emergency dial strings that have to be supported for the current location. As a consequence, location information has to be available even before an emergency call can be placed at all.

A certain PSAP may be responsible for a big area, for example, a city, a county or canton, or a whole federal state. This means that finding the right PSAP does not require very accurate location information, but without location information, it is impossible to find the right PSAP. So location information is again necessary right before the emergency call is set up. Since the Internet does not know any borders and spreads around the globe, a VoIP emergency call without location information is impossible to route (and indistinguishable from a normal call unless there is some kind of red emergency button on the user interface or there is a manual configuration of emergency dial strings). Where should such a call be connected? To a PSAP in the country of the voice provider? Or to the PSAP serving the address the user provided on the sign-up form for the voice provider? What if this entry is no longer up to date or the user is roaming? What if this is an anonymous service, with no pre-registration?

When the connection to a PSAP call taker is established, one of the first questions will be "where is your emergency?". When the caller is able to provide sufficient location information, automatic location information is not that important (however, it could speed up and ease the work of the call taker). Unfortunately there are several cases where callers are unable to describe their location. This may happen for many reasons: The caller might not know his location at all, for example, during a journey in a foreign country. Another reason is more psychological: In emergency situations people easily panic and get nervous, unable to think clearly and consequently unable to describe their location precisely. The most dramatic situation is a caller who is near death, dialing the emergency number with his last ounce of strength and unable to describe his position. Or the communication may break down right before the caller is able to tell the location, maybe due to empty batteries of a mobile phone. One should not forget emergency calls placed by children where it might not be possible to get a specific answer. So it should be clear that location information is important for a high quality emergency call system, helping to get action forces as fast as possible to the emergency situation.

One more thing to point out about automatic location information provided to the PSAP: The fact that PSAPs know callers' location might also deter many fake emergency calls, since the caller would know they can be located and held liable.

When an emergency call is connected to the wrong PSAP (one not serving the location of the caller), valuable time is lost. After the call taker asks the caller for his location, the call has to be manually rerouted to the responsible PSAP. So the caller will probably find himself on hold a second time if no call taker is free at the moment. When the second call taker picks up, he will probably ask the same question as the first one. Sometimes call takers even tell callers to hang up and call the emergency number again when they do not understand why these calls were routed to the wrong PSAP. But in the case of

VoIP emergency calls, this is not helpful. The manual rerouting from one PSAP to another usually only works inside a single country at the moment.

Summing up the importance of location information, location is needed for the following tasks:

- Determining the local emergency dial strings (and thus recognizing that call to these numbers are emergency calls).
- Determining the responsible PSAP.
- Response planning at the PSAP and dispatch of first responders.

So location information is needed before an emergency call can be placed. To point out the importance of location information, consider just two figures from the Cgalies final report (Ludden et al., 2002), published in 2002 by the European Union based on a database of 40 million mobile emergency calls in the EU per year. Because of missing automatic location information, help was significantly delayed for 3.5 million emergency callers. Even more dramatic: for 2.5 million emergency calls, no help could be sent out at all since the caller could not provide location information. These figures show that an improvement of the current emergency call system would be necessary since in any emergency situation every second counts. Automatic location information that is conveyed to the PSAP can be one important component to get responders to an emergency more quickly.

Of course, there are also some emergency calls where the location of the caller is not important at all. For example, when the caller alerts the authorities for a friend having an emergency at a different location. However, in most cases, the location of the caller is valuable and helps to save time when handling the emergency. Even when the caller is able to describe the location, automated location information showing up on the call taker's screen can save a few valuable seconds.

When talking about location determination, accuracy is an important factor. To discover the local emergency dial strings and PSAPs for a caller, a rough determination may be sufficient (e.g., to the state or city level). However, when first responders are looking for the scene of the emergency, more accurate location information is needed. Even one meter accuracy might be desirable (Malenstein et al., 2006, p. 25), for example to determine on which lane of a road an accident has happened, and therefore to direct the ambulance on an optimized route to the scene of the accident. Valuable time can be saved. Knowing the direction the car was driving can also be helpful (Rosen et al., 2010).

Another example is an emergency situation on the bank of a river or canal – action forces have to be sent to the right bank and therefore accurate location information is necessary. Even knowing the position of an injured person with about 50 or even 10 meters accuracy, it might be tough to find a person in a thicket or forest.

Meeting these requirements for precise location information could entail significant investment and costs. However, this should be balanced against the savings to the national economy due to the improved emergency call system (Neumann, 2004). This savings may result from reduced death rates, reduced rehabilitation times, or shorter hospital stays after medical emergencies since the improved emergency call system might allow emergency doctors to arrive sooner on the scene of the emergency and start treatment. Besides these obvious impacts, there are also some side benefits to local and regional economies. For example, accurate location information of road accidents may allow operation centers to redirect traffic or to simply close just one lane of the road by changing appropriate traffic lights. The resulting reduction of traffic jams has a positive impact on national economics.

Of course, there are also commercial applications of location information. Especially as people have come to use mobile devices more heavily, there has been a proliferation of location-based applications for everything from finding the local coffee shop to finding a date. Location services developed to support emergency services could also be used to support commercial location-based applications.

The timely arrival of first responders is advantageous not only for medical emergencies. Just think of the potential reduction of damage when the police react more quickly to robberies, or the fire department to fires.

Even though location information is the key information for emergency calls, it is of course not the sole requirement for emergency calls, as discussed in the previous section.

References

Ludden B, Pickford A, Medland J, Johnson H, Brandon F, Axelsson LE, Viddal-Ervik K, Dorgelo B, Boroski E and Malenstein J (2002) *Report on Implementation Issues Related to Access to Location Information by Emergency Services (E112) in the European Union*.

Malenstein J, Terpstra T, Jaaskelainen M, Medland J and Rooke A (2006) *Position Paper – PSAP expert working group on PSAP eCall requirements*.

NENA IP-Capable PSAP Minimum Operational Requirements Standard (58-001) (2007). National Emergency Numbering Association (US).

Neumann A (2004) *Standortinformationen für Rettungsdienste*.

Rosen B and Polk J (2010) Best Current Practice for Communications Services in Support of Emergency Calling. Internet Draft (work in progress) draft-ietf-ecrit-phonebcp.

Rosen B, Tschofenig H and Dietz U (2010) Best Current Practice for IP-based In-Vehicle Emergency Calls. Internet Draft (work in progress) draft-ietf-ecrit-framework.

3

The ECRIT Emergency Calling Architecture

The Internet Engineering Task Force (IETF) has developed an architecture for VoIP-based emergency calls, the ECRIT (Emergency Context Resolution with Internet Technologies) architecture. Almost all the necessary components are already standardized, just a few details are left to be finalized. In the IETF, a few different working groups – ECRIT, GEOPRIV and to a lesser extent SIPCORE – are working on topics related to emergency calling, developing the framework for the next generation of emergency calls. For VoIP-based emergency calls, the ECRIT architecture mainly focuses on SIP (Session Initiation Protocol), but the basic ideas are also true for other protocols (such as H.323 or XMPP). In this chapter, we'll use SIP messages in the examples, other protocols might be supported in a similar way.

(In particular, the only change to the process that is needed to support a protocol other than SIP is for PSAPs to advertise a new contact URI in LoST. This will allow endpoints that support this new protocol to discover their local PSAP and contact it. SIP has also been extended so that session initiation messages carry the caller's location, and it would be helpful if new protocols had a similar functionality.)

The ECRIT architecture is focused mainly on placing emergency calls via the Internet; improvements to the current emergency calling system for traditional fixed lines or mobile phones is beyond the scope of the IETF. Moreover, strategies for transitioning to Internet-based emergency calling have not been discussed in the IETF, since the specifics of legacy systems vary quite a bit, making it infeasible to develop a global transition strategy.

As there is still some ongoing work within the IETF (see Chapter 8 for a discussion of current topics), this book represents the status at the

VoIP Emergency Calling: Foundations and Practice Karl Heinz Wolf and Richard Barnes

time of writing. As noted above, though, this discussion captures the essential structure of the ECRIT architecture.

For a VoIP user to have consistent access to emergency calling, it is really important that Internet standards for emergency calling via VoIP are developed and implemented. Further, as the Internet knows no borders – more like email rather than traditional telephones – the ECRIT architecture provides a consistent way to reach emergency services from any Internet connection anywhere in the world.

3.1 Overview

To place an emergency call over the Internet, the following information is necessary (additionally to an Internet connection):

Location Information: Location information is the fundamental information for an emergency call. Without location information, an emergency call over the Internet is completely impossible. Location information is necessary to determine the responsible PSAP and to plan a response to the emergency.

Local Emergency dial strings: Emergency calls are detected by a sequence of dialed numbers, so all local emergency dial strings have to be known.

Contact information of the responsible PSAP: In order to connect an emergency call to the responsible PSAP, the PSAP needs to publish its contact information (e.g., a SIP URI). The ECRIT architecture does not necessarily involve an VoIP provider, so that emergency calls can also be directly connected to a PSAP (without using the infrastructure of any VoIP provider). The routing possibilities are discussed in Section 3.4.

Figure 3.1 shows the overall ECRIT emergency calling architecture as well as the sequencing of the individual steps. Additionally, the protocols that are proposed by the IETF are noted in square brackets (these protocols will be discussed in subsequent sections).

When powering on a VoIP device, it will first run the network configuration (step 1). In most networks, this will happen by using a DHCP (Dynamic Host Configuration Protocol) server, which assigns the device an IP address, among other things. The DHCP server can also configure the device with location information or a mapping server address.

Step 2 in Figure 3.1 shows the registration with a VoIP provider. As already noted, this step is not mandatory and could also happen later

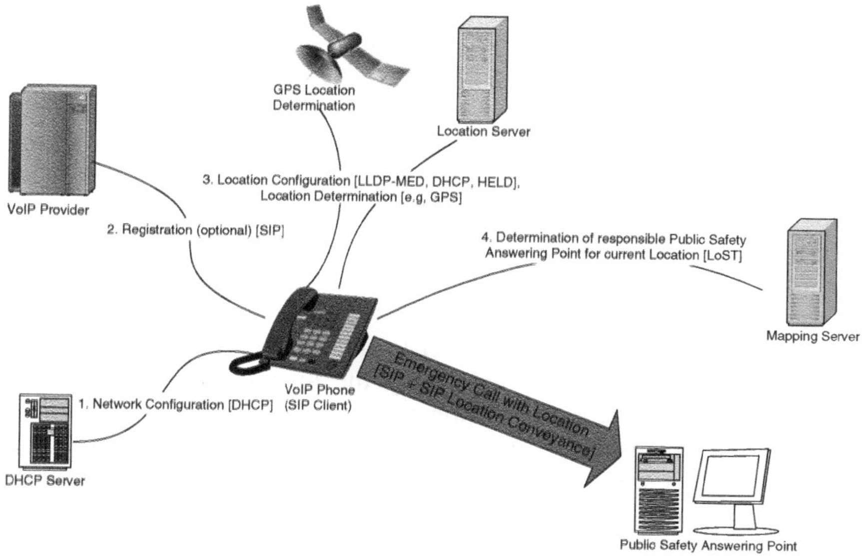

Figure 3.1 The ECRIT emergency calling architecture. Protocols are indicated in square brackets.

on. It is also possible for the VoIP provider to supply the phone with additional configuration data (e.g., with server addresses), but there are not yet any standard mechanisms for such configuration.

The third step is to acquire location information: This can be achieved by either using a protocol for location configuration, like HTTP Enabled Location Delivery (HELD) or DHCP, or by some means of location determination, for example, by a Global Positioning System (GPS) receiver. Since this step is crucial, Chapter 4 discusses location mechanisms in detail, in addition to the discussion in this chapter of how it fits into the general architecture. No matter how location information is acquired, it has to be expressed in a standardized format (see Section 3.2). Without any location information, the subsequent steps cannot be performed and an emergency call is actually impossible.

The next step (step 4) is to contact a mapping server, which provides the mapping between the caller's location and the desired emergency service to the responsible PSAP. The reply from the mapping server contains the contact information of the PSAP (e.g., the SIP URI) and the emergency dial string (e.g., 911). To contact a mapping server, the IETF developed the Location-to-Service Translation Protocol (LoST), which is discussed in Section 3.4.

As soon as the mapping is completed, the phone has gathered all the necessary information and is ready to place an emergency call. Note that all these tasks have to be completed before an emergency call, and not just before the call is connected, but before it is even dialed. It might be necessary to refresh location information and mapping information at the time of an emergency call, but it is essential to perform all these steps when starting up the phone – even though the user might not place an emergency call. The reason for this is quite simple: emergency calls are usually initiated by dialing a well-known short dial string like 911. Hence, the phone has to know that this number is an emergency dial string at its current location in order to treat this call as an emergency call. In order to obtain the dial string, the mapping server has to be contacted. As input for the mapping process, location information is needed (dial strings are a local matter). Consequently, all the above-mentioned steps are necessary to enable the phone to detect and place an emergency call.

Note that a VoIP service provider can actually also detect emergency calls, in addition to the calling device itself. The VoIP provider, however, would have to know the caller's location information and perform all the above steps on behalf of the client, and in most cases, the VoIP provider cannot determine the location of a client (because of the separation of transport and service on the Internet, users log in to the VoIP service from anywhere). A static configuration of all world-wide dial strings is also not a satisfactory solution. Moreover, users might not want to send location information to their VoIP provider for any call to let the provider figure out whether this call is an emergency call or not. Since a VoIP provider is even not mandatory for VoIP, it is preferred that the client detects emergency calls, but the actual emergency call could be connected via the VoIP provider's network (see Section 3.5.2 for more on routing options).

When an emergency call is recognized, the contact information of the responsible PSAP is used to establish a connection over the Internet and the call is specially marked. Moreover, the ECRIT architecture allows location information to be conveyed to where the information is needed: right to the PSAP (see Section 3.5 for more details on the actual emergency call).

3.2 Location Information

In this context, location information is of course information about the physical location of an entity – in this case, an emergency caller.

Location information can be given either as a geographical position (geodetic) or as an address (civic). A uniform document type for location information has been standardized by the IETF: the Presence Information Data Format – Location Object (PIDF-LO). A PIDF-LO document carries location information for an entity as well as presence information and privacy rules for how recipients should handle the object. Location Information itself can be transported either as *by value* (the actual PIDF-LO) or *by reference* (a Uniform Resource Identifier (URI) that refers to a PIDF-LO document). This section also covers conveyance of location information within SIP messages, which is the way that location information is first delivered to a PSAP during an emergency call.

3.2.1 PIDF-LO

PIDF-LO is the IETF's document type for location information, standardized in RFC 4119 (Peterson, 2005) and updated by RFC 5139 (Thomson and Winterbottom, 2008) and RFC 5491 (Winterbottom et al., 2009). As the name implies, PIDF-LO is an extension to the PIDF format (RFC 3863, Sugano et al., 2004), which was created for presence information. Since the privacy requirements for presence and location information are similar, the GEOPRIV working group started developing a unique XML format for location information based on PIDF.

So a PIDF-LO can hold location information in addition to the presence information in the original PIDF format. How location information itself is acquired or how a PIDF-LO is conveyed to a location recipient are not part of the PIDF-LO specification and are discussed in subsequent sections of this book.

Let's first have a look at a basic PIDF-LO document (location information itself is not contained in this first example):

```
<?xml version="1.0" encoding="UTF-8"?>
 <presence xmlns="urn:ietf:params:xml:ns:pidf"
    xmlns:gp="urn:ietf:params:xml:ns:pidf:geopriv10"
    entity="pres:sample@example.com">
  <tuple id="0815">
   <status>
    <gp:geopriv>
      <gp:location-info>
      <!-- location information is inserted here -->
```

```
      </gp:location-info>
      <gp:usage-rules>
        <gp:retransmission-allowed>
          no
        </gp:retransmission-allowed>
        <gp:retention-expiry>
          2010-08-10T09:00:10+02:00
        </gp:retention-expiry>
      </gp:usage-rules>
    </gp:geopriv>
  </status>
  <timestamp>2010-08-10T08:31:00+02:00</timestamp>
 </tuple>
</presence>
```

In PIDF-LO, the element at the root is the "presence" element, with an "entity" attribute that identifies the entity whose presence and location this document describes. The "tuple" element is right below, with a unique random ID. A "tuple" element must hold a "status" element and may optionally have other elements; the "timestamp" element, for example, is especially important for emergency calls to show how fresh the information is. The format for the timestamp is defined in RFC 3339 (Klyne and Newman, 2002), which is a profile of the ISO 8601 format used for time and date elements in XML (*ISO 8601:2000. Data elements and interchange formats – Information interchange – Representation of dates and times (2000)*). The timestamp in the example above is from August 10th 2010, 8:31 Central European Summer Time (UTC + 2 hours). All elements mentioned so far are all defined in the base PIDF specification. The extension for location information in the PIDF-LO (Peterson, 2005) consists of a complex element "geopriv", which contains four important elements: "location-info", "usage-rules", "method" and "provided-by". The "location-info" and "usage-rules" elements are mandatory, the other two are optional.

location-info: This element is mandatory and contains the actual location information, either as geographic location information or as an address. The contents of this element depend on whether the location information is being provided in civic or geodetic form, and will be discussed in greater detail below.

usage-rules: These mandatory usage rules can be used to define how the location recipient should handle this location object:

- Whether the location recipient may share this location information with a third party ("retransmission-allowed").
- How long the location information can be stored ("retention-expires").
- A link to further rules and restrictions ("ruleset-reference").
- Additional human-readable rules ("note-well").

Of course, it may be necessary to disregard some of these preferences during an emergency, but the presence of the rules allows the recipient to know the user's preferences and follow them when possible.

method: This optional element tells which method was used to actually determine the location information (i.e., to figure out where the device is located). Some possible values are "GPS", "RFID", "LLDP-MED" or "Wiremap", among several others. A registry of possible method tokens is maintained by the Internet Assigned Numbers Authority (IANA), available at: `http://www.iana.org/assignments/method-tokens`.

provided-by: This optional element contains XML information on who is responsible for this location information. To ensure interoperability, the XML contents have to be in an IANA-registered XML namespace. At the moment, the only registered namespace is the "dataProvider" namespace (see `http://www.iana.org/protocols/` under "XML Namespaces for 'provided-by' elements for use with PIDF-LO objects"). This element can be used to tell the location recipient who can be contacted in case of problems with the provided location information. The "dataProvider" schema, however, is tailored to the needs of the US; it is essentially a member ID for a NENA member company. In situations where the recipient can directly authenticate the source (e.g., using a digital certificate), this element is dispensable anyway since the identity of the authenticated source can be used (e.g., the "subjectAltName" certificate attribute, as in HTTPS).

So far we've focused on the general structure of a PIDF-LO document, but now let's have a look at how the most important information, the location information, is actually embedded. The element "location-info" may hold geodetic location information based on the Geography Markup Language (GML) by the Open Geospatial Consortium, or civic addresses in a format defined by the IETF.

A sample location (a GML point) in a PIDF-LO is shown below:

```
<?xml version="1.0" encoding="UTF-8"?>
 <presence xmlns="urn:ietf:params:xml:ns:pidf"
    xmlns:gp="urn:ietf:params:xml:ns:pidf:geopriv10"
    xmlns:gml="http://www.opengis.net/gml"
    entity="pres:sample@example.com">
  <tuple id="0851">
   <status>
    <gp:geopriv>
      <gp:location-info>
        <gml:location>
          <gml:Point gml:id="point1"
             srsName="urn:ogc:def:crs:EPSG::4326">
            <gml:pos>48.14 16.94</gml:pos>
          </gml:Point>
        </gml:location>
      </gp:location-info>
      <gp:usage-rules>
        <gp:retransmission-allowed>
           yes
        </gp:retransmission-allowed>
        <gp:retention-expiry>
           2010-08-19T21:02:00Z
        </gp:retention-expiry>
      </gp:usage-rules>
    </gp:geopriv>
   </status>
   <timestamp>2010-08-19T19:42:55Z</timestamp>
  </tuple>
 </presence>
```

This PIDF-LO document describes the point with the coordinates 48.14N 16.94E in the WGS84 system (indicated by the attribute "srsName" with the value "urn:ogc:def:crs:EPSG::4326"). Without going into too much detail of geography, geodetic coordinates have to be defined with respect to some geometric model of the Earth, called a "spatial reference system" or SRS. WGS84 is the most common SRS in use today, mainly because it is the system in which GPS coordinates are expressed. GML requires the SRS for a location to be specified in the "srsName" attribute, but all examples in this book will use the WGS84 SRS.

Actually, GML is a very extensive format that is able to describe different object types. In order to reduce complexity (otherwise every emergency calling client would have to implement a complete GML specification), the RFC 5491 specifies a subset of GML that defines a set of simple shapes (Winterbottom et al., 2009):

- Point: A (latitude, longitude) pair or (latitude, longitude, altitude) triple.
- Polygon: A 2-D region defined by a sequence of 2-D or 3-D points (all at the same altitude), implicitly connected by lines.
- Circle: A 2-D region defined by 2-D center point with a radius.
- Ellipse: A 2-D region defined by 2-D center point with major and minor axes, and an orientation.
- Arc Band: 2-D region defined by a 2-D center point, with inner and outer radii and bounding angles.
- Sphere: A 3-D region defined by 3-D center point with a radius.
- Ellipsoid: A 3-D region defined by a 3-D center point with major, minor, and vertical axes, and an orientation.
- Prism: A 3-D region defined by a polygon (with 3-D points, all at the same altitude) and a height.

These basic shapes, illustrated in Figure 3.2, should be supported by software that receives PIDF-LO objects. PIDF-LO samples are available in RFC 5491.

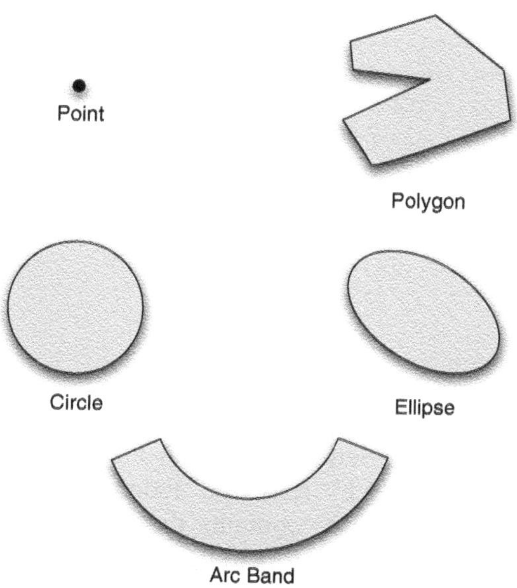

Figure 3.2 2D Shapes for use in PIDF-LO.

When location information is available as a civic address, the PIDF-LO encodes the address as a series of name-value pairs that identify individual address components:

```
<?xml version="1.0" encoding="UTF-8"?>
<locationResponse xmlns="urn:ietf:params:xml:ns:geopriv:held">
 <presence xmlns="urn:ietf:params:xml:ns:pidf"
  xmlns:gp="urn:ietf:params:xml:ns:pidf:geopriv10"
  xmlns:cl="urn:ietf:params:xml:ns:pidf:geopriv10:civicAddr"
  entity="pres:sample@example.com">
  <tuple id="65537">
   <status>
    <gp:geopriv>
      <gp:location-info>
        <cl:civicAddress>
          <cl:country>AT</cl:country>
          <cl:A1>Vienna</cl:A1>
          <cl:A3>Vienna</cl:A3>
          <cl:RD>Karlsplatz</cl:RD>
          <cl:HNO>1</cl:HNO>
        </cl:civicAddress>
      </gp:location-info>
      <gp:usage-rules>
        <gp:retransmission-allowed>
           yes
        </gp:retransmission-allowed>
        <gp:retention-expiry>
           2010-01-11T04:07:00Z
        </gp:retention-expiry>
      </gp:usage-rules>
    </gp:geopriv>
   </status>
   <timestamp>2010-01-01T10:10:10Z</timestamp>
  </tuple>
 </presence>
</locationResponse>
```

This sample shows an address in Vienna, Austria. The element "country" holds the two-letter ISO 3166-1 country code "AT" for Austria. The element "A1" tells the federal state and "A3" the commune. The element "RD" contains the road name (Karlsplatz in this example). The house number is to be found in the element "HNO". Note that all element names are written in uppercase letters – except the "country" element. A registry of allowed address elements for

PIDF-LO is maintained by IANA and available at `http://www.iana.org/protocols/` under "CAtypes". These elements are sometimes referred to as CAtypes, an abbreviation for "Civic Address Types", which is the name for the analogous elements of the DHCP civic address format (Schulzrinne, 2006) (see Section 4.1.2).

The IETF has simplified the extensive GML specification by restricting the usage to certain profiles. Unfortunately, civic location information (addresses) can be complex, too. With addresses, the problem is different: while it is possible to express coordinates worldwide in a uniform format, the format of addresses differs from country to country, with local idiosyncrasies. For example, many Chinese addresses use a "sub-branch road name", but most US and European addresses have no analogous concept. The IETF has defined a basic set of address elements. However, how they should be used for local addresses of individual countries is not part of the specification. The IETF does not develop country-specific variations of their protocols, and a PIDF-LO specification with all address elements of every country in the whole world seems unrealistic. Hence, the IETF tackles this problem differently: Each country is encouraged to create a "civic address considerations" document that states how the standardized IETF civic address elements should be used for a particular country. For example, Austrian house numbers have a more complex format than the PIDF-LO elements could hold – just the house number part of an address can be "1a - 5a Block 1b Haus 2c Stiege 1" – and PIDF-LO just has a single field ("HNO" element) and a suffix ("HNS" element). For formal usage of PIDF-LO, as for emergency calling, interoperability is an issue. PIDF-LO documents are typically created by someone other than the entity that interprets them (e.g., a network operator creates a PIDF-LO and a PSAP receives it during an emergency call). PSAP operators certainly want to find civic address parts in the same PIDF-LO element, no matter who has created it. Misinterpretation of an address of an emergency situation could have catastrophic consequences.

Since civic addresses are a local matter, the meaning of the PIDF-LO elements can also be arranged locally. And emergency calls should anyway just be connected to a PSAP in the country to which the address consideration applies (or if not, at least the country element shows to which country this call has to be routed).

The IETF GEOPRIV working group published a set of guidelines for civic address considerations in RFC 5774 (Wolf and Mayrhofer, 2010). This document describes general guidelines for mapping a national address structure to the general GEOPRIV civic address format and registering these mapping with IANA. It also includes the first such mapping document, which describes how Austrian addresses should be encoded. To support interoperability, civic address considerations documents are listed in the IANA registry at `http://www.iana.org/assignments/` under "PIDF-LO Civic Address Considerations Registry", so that the recipient of a PIDF-LO can locate them easily in order to interpret a civic address. Hopefully, every country in the world will create such a document soon.

When a new civic address considerations document is created, it has to be published somewhere on a public website, preferably on the web-page of an addressing agency. Note that the civic address consideration documents are not RFCs and are therefore not published by the IETF. Moreover, RFC format restrictions do not apply to all these documents. Hence, they can be PDF or HTML documents, including tables and pictures. Their aim is to clarify how to handle civic addresses in PIDF-LO. So when this document is ready, it will undergo a brief expert review by the IETF to make sure that it meets all the requirements from the guidelines in RFC 5774. Then the document can be put in the IANA registry. As long as the status indicated in the registry is "active", the considerations stay valid. The authors of the considerations document can also request that the status of the document is set obsolete (e.g., when another considerations document is updated and the old one is no longer valid).

When requesting registration of a document, URIs are referenced in the following way, using the IANA XML format:

```
<xref type="uri"
  data="http://www.example.com/addressing/pidflo/"/>
```

The PIDF-LO specification as well as the rules on how to use the address elements is essential for the further steps of an emergency call. Another fundamental concept is the distinction between location by value and location by reference.

3.2.2 Location by Value and Location by Reference

PIDF-LO documents hold location information. When sharing location information with another entity, one has two options: send the PIDF-LO directly to the recipient or send them just a reference to that PIDF-LO. The direct transportation of a PIDF-LO document is called *location by value*, since the value of the location information itself is transported.

Table 3.1 Comparison of location by value and location by reference

Location by Value	Location by Reference
Client has to send the whole PIDF-LO	Client just sends the reference (URI)
Client has to send location updates when location changes	The receiver is able to refresh location information autonomously by dereferencing the URI
Suited for self-determined location information (e.g., GPS), no Location Server needed	A Location Server is necessary, self-determined location information would have to be provisioned on the Location Server (not part of the standardized location configuration protocols) or the phone would have to implement a Location Server itself
No location updates after disconnect	Even when the client's phone hangs up or exits, the location information can still be fetched from the Location Server
Required for the LoST protocol	Not allowed in LoST
Supported by every location configuration protocol	Not all location configuration protocols support location by reference
Location information is sent directly to the receiver	Sender doesn't know whether the receiver is actually able to resolve the reference

Cases where a reference is used (pointing to the actual PIDF-LO on a Location Server) are called *location by reference*. Both options have pros and cons, their characteristics are compared in Table 3.1.

In order to illustrate the way location by value and location by reference work, both options are shown in Figure 3.3 and Figure 3.4. These figures assume that the phone uses a Location Server, for example, with the HELD protocol as discussed in Section 4.1.1. For the location by value case, the phone could of course use some local determination, like GPS, but a Location Server is required for location by reference.

In most cases, the recipient of a location URI can get updated location information simply by using the URI to get the location information from the Location Server. The caller doesn't have to be involved in this process, so it saves on bandwidth and power compared to sending location updates by value. These savings can be especially

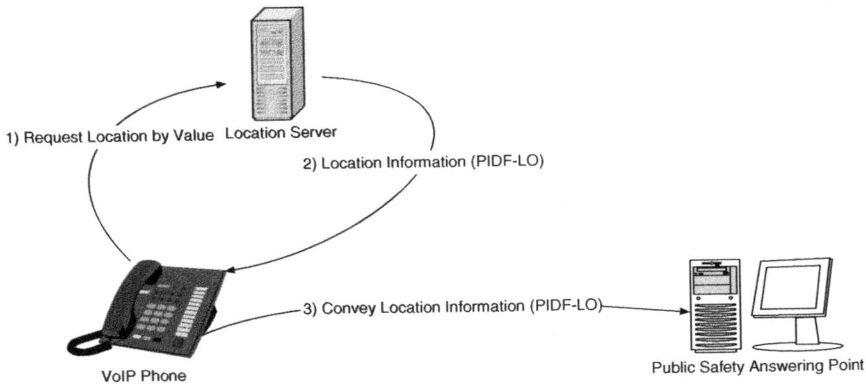

Figure 3.3 Location by value.

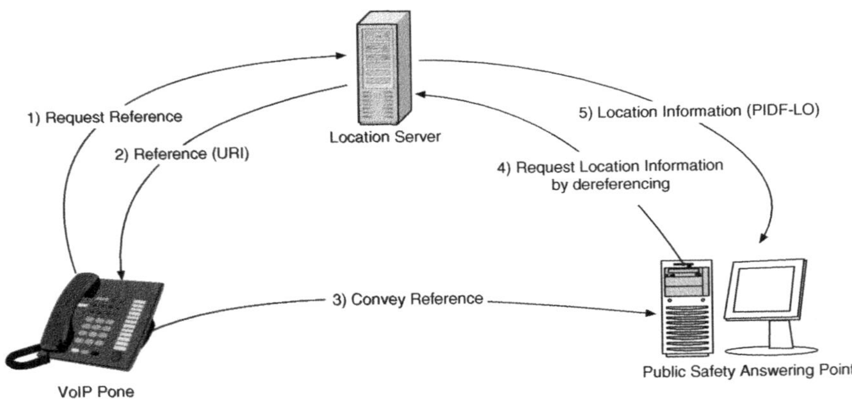

Figure 3.4 Location by Reference.

important for mobile devices. In emergency calls, it can be advantageous to be able to update location information even when the phone call is disconnected – another advantage of not requiring the caller to be involved. (The caller's privacy can still be protected, though: Privacy Rules at the Location Server can be enforced and the URI has just a certain lifetime.) The exception to this pattern arises with so-called snapshot URIs, which will always return the same location information, no matter if it has changed or not. Instead of a stable reference to the caller's current location, a snapshot URI is simply a snapshot of the location information for a particular point in time. In the special case of snapshot URIs, location by reference has mostly the same properties as location by value, but in other cases, location by reference allows the recipient of the reference to easily obtain updated location information.

There are also some cases where it is advantageous to use location by value and location by reference at the same time. For instance, the Location Server can provide coarse-grained location information by value alongside a reference to more precise location information (see Section 4.4 for more details).

3.2.3 Location Conveyance

In a VoIP emergency call, the PSAP cannot necessarily find the caller's location on its own, so the caller needs to send location information to the PSAP as part of his emergency call signaling. The transportation of location information to a location recipient is called location conveyance. Location information can be conveyed either by value as a PIDF-LO or by reference as a URI that points to a PIDF-LO. This section describes conveyance when using the SIP protocol for VoIP signaling, as required by the ECRIT architecture. For other signaling protocols a similar approach might be applicable. In any case, because location information is critical for emergency calls, it has to be ensured that location information for emergency calls is automatically conveyed to the PSAP.

Location information is embedded in a SIP message by way of the Geolocation header. The purpose of this header is to identify where the location information being conveyed can be found. In the case of location by value, the PIDF-LO document is embedded as a body part in a multi-part MIME body of the SIP message, and the Geolocation header contains a pointer to the part containing the PIDF-LO. This pointer is a Content Identification (CID) URI, as described in RFC 2392 (Levinson, 1998). In case of location by reference, the Geolocation header contains an URI pointing to the PIDF-LO outside this message. (The use of Data URIs (see Masinter, 1998) is not recommended, since they would create overly large headers containing a whole location object.)

The following SIP INVITE message conveys location information by value:

```
INVITE sip:vienna-police@10.10.0.46 SIP/2.0
Route: <sip:nic.at;lr>
Via: SIP/2.0/TCP 10.10.0.47:5060;rport;branch=z9hG4bK4r1id0
```

```
To: "Vienna Police" <sip:vienna-police@10.10.0.46>
From: John Doe <sip:jd@example.com>;tag=1nexm5g
Call-ID: c3861f9d-9b78-48e0-916a-272066ebcae7@example
CSeq: 2 INVITE
Max-Forwards: 70
User-Agent: zap/0.2.3
Supported: path
Supported: gruu
Contact: John Doe
  <sip:jd@10.10.0.47;grid=96b9605d45afc2c07500ee4b587>
Supported: geolocation
Geolocation: <cid:pidflo@zap>;inserted-by=10.10.0.47;
  recipient=endpoint;used-for-routing
Content-Type: multi-part/mixed;boundary=boundary1
Content-Length: 1259

--boundary1
Content-Type: application/sdp

v=0
o=zap 0 0 IN IP4 10.10.0.47
s=
c=IN IP4 10.10.0.47
t=0 0
a=ice-pwd:b2n6h3m5xt9yvqhpnv100jil4
m=audio 49152 RTP/AVP 0 8 97 101
a=candidate:1o60zvg 1 UDP 0.5 10.10.0.47 49152
a=candidate:1o60zvg 2 UDP 0.5 10.10.0.47 49153
a=rtpmap:0 PCMU/8000
a=rtpmap:8 PCMA/8000
a=rtpmap:97 speex/8000
a=rtpmap:101 telephone-event/8000
a=fmtp:101 0-15

--boundary1
Content-Type: application/pidf+xml
Content-ID: <pidflo@zap>

<presence xmlns="urn:ietf:params:xml:ns:pidf"
  xmlns:gp="urn:ietf:params:xml:ns:pidf:geopriv10"
  xmlns:gml="http://www.opengis.net/gml"
  entity="pres:geotarget@example.com">
  <tuple id="sg89ae">
   <status>
```

```
      <gp:geopriv>
        <gp:location-info>
          <gml:location>
            <gml:Point gml:id="point1"
                srsName="urn:ogc:def:crs:EPSG::4326">
              <gml:pos>48.201148 16.36954</gml:pos>
            </gml:Point>
          </gml:location>
        </gp:location-info>
        <gp:usage-rules>
          <gp:retransmission-allowed>
            no
          </gp:retransmission-allowed>
          <gp:retention-expiry>
            2011-06-23T04:57:29Z
          </gp:retention-expiry>
        </gp:usage-rules>
      </gp:geopriv>
    </status>
    <timestamp>2011-06-22T20:57:29Z</timestamp>
  </tuple>
</presence>
--boundary1--
```

This SIP message was generated by the Zap! emergency calling SIP client, which is discussed in detail in the practical part of this book (Chapter 6). In the header part of the SIP message (above the first empty line) there are two important lines: the Supported and the Geolocation header. With the Supported header, the SIP client communicates that he understands the concept of location conveyance:

```
Supported: geolocation
```

The Geolocation header points to the actual location information, in this case:

```
Geolocation: <cid:pidflo@zap>;inserted-by=10.10.0.46;
   recipient=endpoint;used-for-routing
```

The URI scheme "cid:" indicates that the PIDF-LO is to be found in the body of the SIP message, which means location by value is being used. The "inserted-by" header parameter indicates which host has inserted this location information. The parameter "recipient" is used to tell for whom this location information is intended (in the example above for the endpoint of the connection). The "used-for-routing" header parameter is included when the location information was used to route the SIP message, for example, to determine the responsible PSAP for an emergency call based on this location information, as discussed in Section 3.4. The number and definition of the header parameters are not yet fixed by the IETF, and every new version of the SIP location conveyance draft has some changes on this topic.

One important note on header usage: If a SIP client supports location conveyance, it has to indicate that fact with an appropriate Supported header. If there is no location information available or the client does not want to convey it, a Geolocation header must not be added. Because of the Supported header, the recipient (e.g., the PSAP) knows that the client supports location conveyance but location information is unavailable or withheld, as opposed to a client who doesn't understand the concept of location conveyance.

Location by value requires that the location information is contained in the body of a SIP message. Usually, the body of a SIP INVITE message contains the Session Description Protocol (SDP) (Handley et al., 2006). In order to accommodate both parts, the session description (a SDP) and location information (a PIDF-LO), a multi-part body is necessary. This is declared with the Content-Type header:

```
Content-Type: multi-part/mixed;boundary=boundary1
```

The body parts are separated by the value of the parameter boundary (in the example above, "boundary1").

The first body part of the example SIP INVITE message holds the session description, followed by a PIDF-LO in the second body part. Every body part starts with headers, giving information about the content.

```
Content-Type: application/pidf+xml
Content-ID: <pidflo@zap>
```

The headers for the PIDF-LO body part indicate that it is in fact a PIDF-LO document ("application/pidf+xml"), and specify the Content-ID for the body part. The same Content-ID is to be found as in the

Geolocation header above – the recipient uses this linkage to find the body part with the location in it. After an empty line the PIDF-LO starts and is constructed as discussed in Section 3.2.1.

For location by reference, a general URI is used. For example, the Geolocation header could look like:

```
Geolocation: <https://lis.example.com/3sfk39kd8as02>;
  inserted-by=192.0.2.1;recipient=endpoint
```

The recipient of the Geolocation header can dereference the URI to get a PIDF-LO. For location by reference it is not necessary to use multi-part SIP messages, just a single (non-multi-part) SDP body. Consequently, the complexity of the SIP messages is reduced, which can be advantageous since not all SIP implementations support multi-part SIP messages. However, with location by reference, there is the potential danger that the recipient (e.g., the PSAP) cannot dereference the URI.

Figure 3.5 shows how a call can be sent to a PSAP including location by reference. In this case, a SIP/SIPS URI is sent in the Geolocation

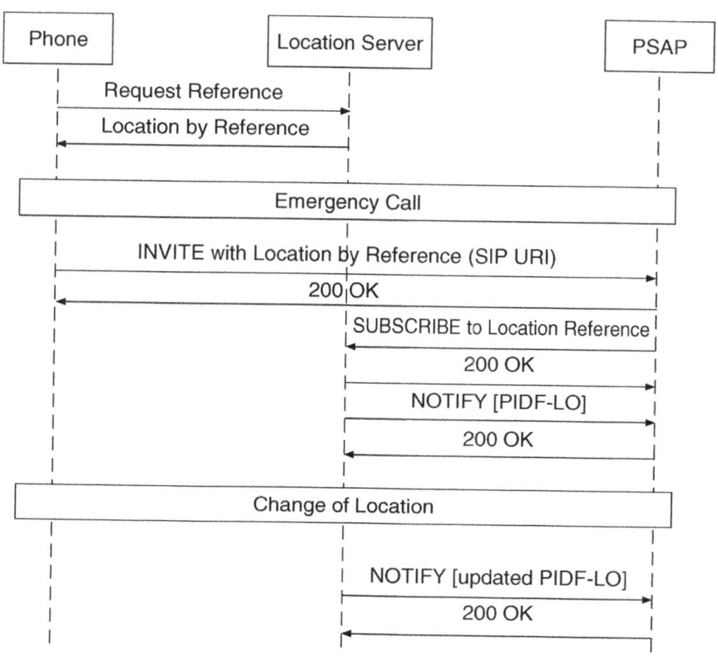

Figure 3.5 Emergency call providing location by reference with a SIP URI.

header. By making use of the SIP event notification mechanism, the PSAP subscribes to the caller's location information at the Location Server (with a SUBSCRIBE message). With the first NOTIFY message the PSAP learns the caller's current location (sent as a PIDF-LO document). Whenever the Location Server has an updated location information, the PSAP will automatically be notified by further NOTIFY messages. If the location reference provided were an HTTP URI, the flow would be similar, except that the PSAP would have to request a location update each time it required one, instead of subscribing to updates and receiving notifications.

If a SIP proxy server wants to add location information (e.g., because the SIP client doesn't support location and the proxy server is able to figure out the location information of the client), it can use only location by reference. This restriction is due to the rule that a SIP proxy server must not modify the body of a message, but may add additional headers. For location by value, the proxy would have to add a Geolocation header and also modify the body to add a body part. So, the only possibility is to add a Geolocation header with a reference to a Location Server holding the location information. Also note that the proxy server must not modify existing location information, which means it is not allowed to alter existing Geolocation headers, but may add an additional header. When adding multiple Geolocation headers, there is the risk of confusing the receiver with different location information. So it's important to mark the location information that was used for routing the call (with the "used-for-routing" header parameter). Otherwise a proxy might try to find out the responsible PSAP with another location information, resulting in a different PSAP and an incorrectly routed call. In cases where the network operator and the VoIP provider are different entities, the proxy server might not be able to add location information anyway.

Again it should be noted that a SIP proxy is not mandatory for a successful emergency call in the ECRIT architecture, and whenever the client has location information, it should be included in the call set-up when placing an emergency call. However, there is a possibility that the client will try to lie about his current location. For example, a malicious caller forges his location to send calls to a far-away PSAP, as part of a denial-of-service attack on the PSAP. Chapter 7 discusses some approaches to mitigate this problem, including signing location objects for location by value and authenticated dereferenced transactions for location by reference.

3.3 Service URNs

The most common emergency services are of course the ambulance, fire brigade and police. In addition to these services, several countries have other services, such as mountain rescue, response to gas leaks, coastguard, poison control, or even forest fire response, to mention just a few.

So in the current system, every country has its own set of several emergency services. In order to identify them in a uniform way, the IETF has created a registry of "Service URNs" (Schulzrinne, 2008). These uniform resource names (URNs) are short text strings that identify an emergency service – independent of the dial string used to reach it. Thus, for example, the fire brigade is internationally identified by "urn:service:sos.fire". All emergency services that are currently defined by the IETF are listed in Table 3.2. Alongside these "sos" Service URNs is a set of URNs for "counseling" services, shown in Table 3.3. The authoritative registry for all of these URNs, maintained by IANA,

Table 3.2 SOS Service URNs

URN	Service
urn:service:sos	Emergency services (generally)
urn:service:sos.ambulance	Ambulance service
urn:service:sos.animal-control	Animal control
urn:service:sos.fire	Fire brigade
urn:service:sos.gas	Gas leaks and gas emergencies
urn:service:sos.marine	Maritime search and rescue
urn:service:sos.mountain	Mountain rescue
urn:service:sos.physician	Medical information
urn:service:sos.poison	Poison control
urn:service:sos.police	Police or other law enforcement

Table 3.3 Counseling Service URNs

URN	Service
urn:service:counseling	Counseling services (generally)
urn:service:counseling.children	Counseling for children
urn:service:counseling.mental-health	Mental health counseling
urn:service:counseling.suicide	Suicide prevention hotline

is available at `http://www.iana.org/protocols/` under "Service URN Labels".

It is almost certain, however, that the current list of Service URNs does not cover all available emergency services worldwide. For example, there is no registration for an emergency service for forest fires (which have their own emergency service in Greece, Italy, and Portugal) or electrical emergencies (Ireland, Egypt, Israel, and Indonesia). Furthermore, some countries also have multiple different police units; this distinction cannot be made with the current set of Service URNs.

In conclusion, it is important to mention that these Service URNs are not intended to be visible to users at all. Users will continue the dial the well-known and easy-to-remember dial strings like 911 or 112. At present, no one is expected to dial a Service URN, like urn:service:sos.police in an emergency situation (even though this expectation might change over time). However, the protocols in the background will make extensive use of the Service URNs. For example, the process for figuring out the emergency dial string and the PSAP contact information for a particular emergency Service URN is discussed in the next section.

3.4 Determining the Appropriate PSAP – the LoST Protocol

The important step of determining the responsible PSAP for a call has to be performed before any emergency call can be placed. This process is essentially a mapping of location information to the responsible PSAP. For this purpose, the IETF developed the Location-to-Service Translation protocol, or LoST (RFC 5222, Hardie et al., 2008). LoST is an XML-based protocol that is able to do more than just tell the client the right PSAP. LoST can also return available services, provide areas for which this mapping is valid, and even validate civic addresses. At the end of this section we also briefly explain how a device can discover a local LoST server.

Of course, all of the XML messages in the LoST protocol need to be carried in some transport protocol in order to get between the client and the server. The LoST specification requires the use of HTTP or HTTPS, largely because using these protocols makes it easy to implement clients and servers (e.g., as CGI scripts or Java servelets).

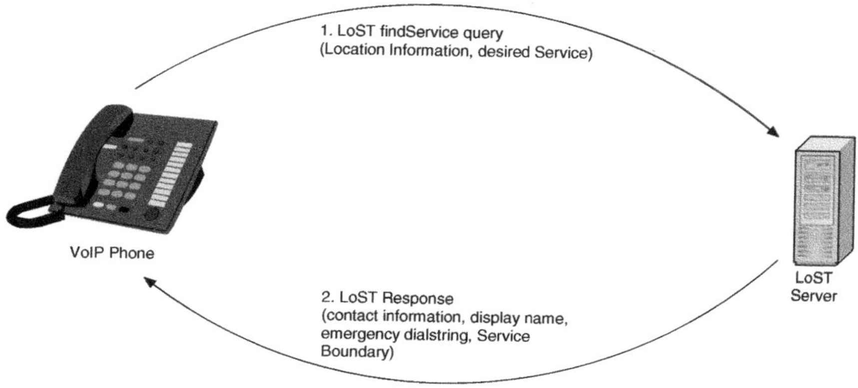

Figure 3.6 The mapping process.

3.4.1 The Mapping Process – findService

Mapping location information to PSAP contact information is the main purpose of the LoST protocol. A LoST query carries two essential pieces of information (see also Figure 3.6):

- Location information for the caller.
- Requested emergency service as Service URN (e.g., urn:service:sos.police).

The LoST server then returns the following output values:

- Contact information of the PSAP (e.g., the SIP URI).
- Display name (e.g., Vienna Police).
- Dial string (e.g., 133 is the emergency dial string for the police in Austria).
- Service boundary (the area for which this mapping is valid, e.g., Vienna).

Let's look at an example of how the mapping process is done via LoST. The caller's phone might send out the following request:

```
<?xml version="1.0" encoding="UTF-8"?>
<findService
  xmlns="urn:ietf:params:xml:ns:lost1"
```

```
  xmlns:gml="http://www.opengis.net/gml"
  serviceBoundary="reference"
  recursive="true">
  <location id="0815b" profile="geodetic-2d">
    <gml:Point id="point1"
          srsName="urn:ogc:def:crs:EPSG::4326">
      <gml:pos>48.201148 16.36954</p2:pos>
    </gml:Point>
  </location>
  <service>urn:service:sos.police</service>
</findService>
```

In this request, the phone wants to find out the responsible police PSAP at the given location. In this case, location information is provided as coordinates. In LoST, the location profile geodetic-2d is used when location is provided as a pair of coordinates or another two-dimensional shape, like a polygon describing the outline of a building. So it's not a whole PIDF-LO document that is sent to the LoST server; just the essential information is extracted by the client in order to create a LoST request. There might be ambiguous results if something other than a point is provided. For example, location information provided as a circle may overlap with the jurisdictions of several PSAPs. In such a case, the LoST server may return all the PSAPs that come into question, or it may choose one PSAP (since the client probably cannot make a better choice than the server about which PSAP to call in case of an emergency call). Actually, in certain areas, PSAP jurisdiction can overlap. So even for a point, the mapping can be ambiguous.

The response to the example request above would look something like the following:

```
<?xml version="1.0" encoding="UTF-8"?>
<findServiceResponse
  xmlns="urn:ietf:params:xml:ns:lost1">
  <mapping
    expires="2011-01-01T01:01:01Z"
    lastUpdated="2008-08-01T20:00:00Z"
    source="authoritative.example"
    sourceId="1234567890">
    <displayName xml:lang="en">
      Vienna Police
    </displayName>
```

```
    <service>urn:service:sos.police</service>
    <serviceBoundaryReference
      source="authoritative.example"
      key="08150815081508150815" />
    <uri>sip:vienna-police@example.com</uri>
    <serviceNumber>133</serviceNumber>
  </mapping>
  <path>
    <via source="resolver.example"/>
    <via source="authoritative.example"/>
  </path>
  <locationUsed id="0815b"/>
</findServiceResponse>
```

This findServiceResponse gives the mapping information, along with a few other helpful things. The essential elements are the “uri”, containing the contact information of the PSAP, and the “serviceNumber” element, containing the emergency dial string. This is the information that is required for an emergency call: The emergency call can be detected using the dial string, and routed using the URI.

The display name can be shown to the user for information purposes.

The area for which the mapping is valid was returned as a reference in the example (the “serviceBoundaryReference” element). The caller’s phone could use the information in this reference to download the geographical outline of the PSAP’s service area. Knowing this area is useful for mobile devices and is discussed in Section 3.4.4.

It is important to mention that the “serviceNumber” element is defined as optional in LoST. The reason for this is simple: the IETF wanted to allow this protocol to be used for other, non-emergency, services as well as emergency services. A pizza service, for example, might not necessarily have a special dial string that could be put in the “serviceNumber” element. However, for emergency service usage, this element is mandatory and very important (even though it was forgotten in some emergency call examples contained in early IETF drafts!). The topic detection of emergency calls is covered in Section 3.5.1.

By including the “locationUsed” element, the server tells the client which location information was used for the mapping process. So if the client sent more than one type of location, he knows which profile was used for the mapping.

3.4.2 Retrieving the Service List – listServicesByLocation

So far, we have only discussed the mapping process for a single emergency service. But how does a client actually know which services are available? One might argue that the list of possible Service URNs is defined and could be programmed into the client. However, the list can change, and an update to this service list could be difficult to deploy. Moreover, it might not be practical to query all defined services, even though just a few of them exist at a particular location. For instance, in the US, where there is typically only one service ("urn:service:sos") reachable by 9-1-1, searching for all the possible Service URNs would entail a great deal more effort. To resolve these ambiguities, LoST has another query, called "listServicesByLocation", which lists all available services at a particular location:

```
<?xml version="1.0" encoding="UTF-8"?>
<listServicesByLocation
  xmlns="urn:ietf:params:xml:ns:lost1"
  xmlns:gml="http://www.opengis.net/gml"
  recursive="true">
  <location id="0815" profile="geodetic-2d">
    <gml:Point srsName="urn:ogc:def:crs:EPSG::4326">
      <gml:pos>48.201148 16.36954</p2:pos>
    </gml:Point>
  </location>
  <service>urn:service:sos</service>
</listServicesByLocation>
```

This query asks for a list of the emergency services (see the service element) that are available for the given location (provided as coordinates). A response to this query would have the following form:

```
<?xml version="1.0" encoding="UTF-8"?>
<listServicesByLocationResponse
  xmlns="urn:ietf:params:xml:ns:lost1">
  <serviceList>
    urn:service:sos.ambulance
    urn:service:sos.fire
    urn:service:sos.gas
    urn:service:sos.police
  </serviceList>
  <path>
```

```
    <via source="resolver.example"/>
    <via source="authoritative.example"/>
  </path>
  <locationUsed id="0815"/>
</listServicesByLocationResponse>
```

In this example, there are ambulance, fire, gas and police services for the client's location. After receiving this response, the client can perform a mapping for these services including the respective top-level service itself (with "findService" queries as described above). Moreover, the client now knows that it doesn't have to do any mapping queries for services other than the ones in the "serviceList" element.

It is important to point out that the top-level service itself is not contained in the service list – only immediate child services are. Hence there is a little pitfall: one should not forget to perform the mapping for the top-level service itself even though it does not show up in the response.

One might wonder how to figure out the available top-level services then. The top-level services for emergency services can be stored in the implementation (urn:service:sos and urn:service:counseling), but other services may evolve and a hard coded list of services is not feasible to maintain. For this reason a client can place a "listServicesByLocation" query without a "service" element in the request and the server will return available top-level services in the response. In order to figure out which child services are available below, the client has to issue "listServicesByLocation" requests for these top-level services. Finally, a "findService" request results in the desired mapping information.

Also note that there is another query defined in LoST, namely listServices. This query is rather for diagnostic purposes only since it does not include any location information. The server only returns services it understands and not the services that are available at the location of the client as it is the case with the "listServicesByLocation" query.

Combining the queries explained above, the complete mapping process has two steps:

1. "listServicesByLocation" query/queries to get the available services at the current location.
2. "findService" queries for each of the services returned in the "serviceList".

It is also possible for a SIP proxy to do the mapping on behalf of the SIP client. However, the client would have to send location information with the call, so that the proxy is able to perform the LoST query in order to determine if the called number actually is an emergency dial string at the location of the caller or not. A static list of emergency dial strings or just the emergency dial string at the location of the proxy server might not be enough in most cases, since there are many different dial strings in use around the world. Thus, the preferred practice is for the client to perform the mapping itself.

3.4.3 Address Validation

LoST can also be used to validate civic addresses. Civic addresses can be thought of as keys into a database of location maintained by some authority, such as a postal service or building registry. (In fact, many jurisdictions do maintain such databases and use them for geocoding.) The goal of civic address validation is to ensure that a given civic address is a valid key for the database. For example, in the US, addresses are kept in a Master Street Address Guide (MSAG) for a jurisdiction, so address validation would act as a check that the address is present in the MSAG. Civic address validation is more than a check on data-entry errors; it can be valuable even when an address is correctly entered, since there are often multiple ways to express a given address, not all of which are valid for emergency services.

Within LoST, the client can request civic address validation by setting the optional attribute "validateLocation" with the value "true" when performing a "findService" query. As a result, the client will get back a "locationValidation" element, as shown in the following example:

```
<locationValidation>
  <valid>country A1 A3</valid>
  <invalid>RD</invalid>
  <unchecked>HNO</unchecked>
</locationValidation>
```

The result of this validation tells that the elements country, A1, and A3 are valid. However, the road name (RD element), was invalid and the house number (HNO element) was not checked by the LoST server. The client can provide this information to a user to help correct any errors that are found in a civic address.

3.4.4 *Areas of Responsibility – serviceBoundary*

Emergency call centers are typically responsible for a prescribed geographical area. Because of this fact, one might think that a mobile user would have to perform the mapping process after every movement to ensure that its list of responsible PSAPs was up to date. Especially for phones with built-in GPS receivers giving continuous location updates, this could generate a lot of traffic and additional load on the LoST infrastructure, and quickly drain the device's battery. However, when there is just a small change in location, it is very likely that still the same emergency call center is responsible. But what is a small change in location? How does the device know to perform a mapping query again without missing a change?

These questions are answered by the so-called "serviceBoundary" element in LoST. A service boundary describes to the client the area that a PSAP is responsible for. As long as a client just moves around within this area, no new mapping is necessary (see Figure 3.7). So this concept helps to reduce the number of queries a mobile device has to do when moving. Note that each emergency service might have its own different area of responsibility (e.g., mountain rescue near mountains, coastguard near water), so the service boundary is bound to a particular service.

A service boundary may be requested "by value" or "by reference" (these terms are similar to the corresponding terms for location information, see Section 3.2.2). The "serviceBoundary" attribute in the example

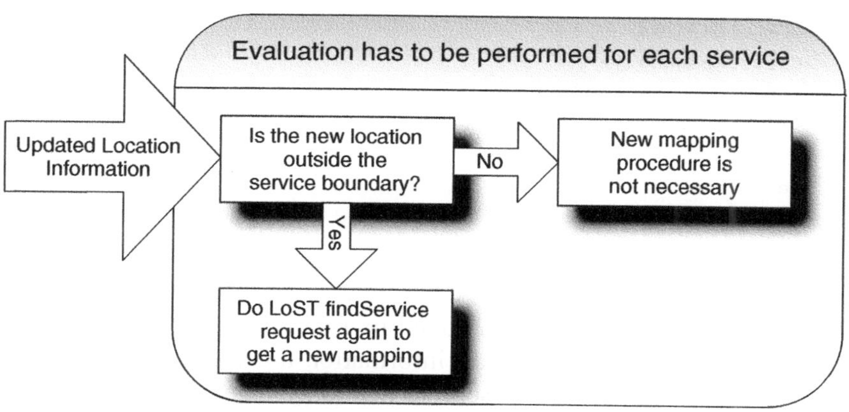

Figure 3.7 Determining whether updated mappings are required.

at the beginning of Section 3.4.1 is set to “reference”, meaning that the client would like the service boundary to be provided by reference. The response to this query contains a “serviceBoundaryReference” element, with a source and a key for the reference. To obtain the service boundary itself, the client sends the key to the source. That is, it sends a second query to the server specified in the “source” attribute, which requests the service boundary using the key:

```
<?xml version="1.0" encoding="UTF-8"?>
<getServiceBoundary
 xmlns="urn:ietf:params:xml:ns:lost1"
 key="08150815081508150815"
 />
```

In response, the server returns the service boundary for the PSAP in question:

```
<?xml version="1.0" encoding="UTF-8"?>
<getServiceBoundaryResponse
  xmlns="urn:ietf:params:xml:ns:lost1">
  <serviceBoundary profile="civic">
    <civicAddress xml:lang="de"
      xmlns="urn:ietf:params:xml:ns:pidf:geopriv10:civicAddr">
      <country>AT</country>
      <A1>Vienna</A1>
    </civicAddress>
  </serviceBoundary>
  <path>
    <via source="resolver.example"/>
    <via source="authoritative.example"/>
  </path>
</getServiceBoundaryResponse>
```

In this case, the emergency call center is responsible for Vienna, Austria. Since a key is always bound to a particular service boundary, a client may cache the boundary information together with the key, so it won’t have to retrieve the service boundary again if it gets another serviceBoundaryReference with the same source and key. Whenever the service area for a PSAP changes, the new area will also have a new key assigned.

Whenever a client wants to get the service boundary by value directly (in the mapping response), the “serviceBoundary” attribute in the “findService” request has to be set to “value”. Then the “findServiceResponse” will contain a “serviceBoundary” element, in the same format as shown in the “getServiceBoundaryResponse” example above. Clients that do not want to evaluate the service boundary (e.g., fixed VoIP phones) should request the boundary by reference and not make a request to dereference it. However, mobile clients in particular may benefit from evaluating the service boundary.

There is also another boundary information for LoST, namely the service list boundary, which informs a client about the area a service list is valid for. This LoST extension is useful to ensure that a mobile client does not miss a change in available services. This issue is discussed in Section 8.2.5.

3.4.5 LoST Server Discovery

In order to request mapping information, the client has first to know the address of a LoST server. One might imagine that such an address could be manually configured, since a LoST server may redirect queries it cannot process itself. However, it saves valuable time and bandwidth to directly contact the right LoST server, namely the one that has mapping data for the user’s current location.

So the IETF also worked out a possibility to automatically configure the address of a LoST server, based on DHCP and the Domain Name System (DNS) (Schulzrinne et al., 2008). In order to direct clients to the appropriate LoST server, an ISP configures its DHCP server to send out a special “LoST discovery” option, which contains a domain name. When a client receives this option, it uses the domain name in a DNS Naming Authority Pointer (NAPTR) query to find the URI for the LoST server that it should contact. NAPTR records for advertising LoST server URIs have the following form:

```
example.com  IN NAPTR 100 10 "u" "LoST:https" \
           "!.*!https://lost.example.com/!"  ""
```

Summing up, the whole process of mapping has to start with LoST server discovery first, of course. A typical sequence is shown in Figure 3.8.

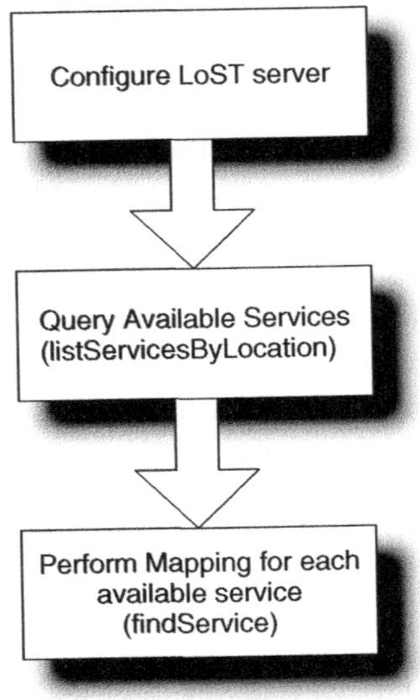

Figure 3.8 Determining the available emergency services and the corresponding PSAPs using the LoST protocol.

3.4.6 LoST Architecture

So far we have discussed the LoST protocol itself and procedures that a client follows. However, one might wonder how the LoST servers are organized and how the global mapping infrastructure is supposed to work. LoST information is like DNS information, in that it exists in a globally distributed database (Schulzrinne, 2009). Figure 3.9 provides an overview of the LoST architecture.

The LoST infrastructure consists of the following basic elements:

Seeker: A seeker is the client (e.g., a VoIP phone) who uses LoST servers, most commonly with the help of a resolver to obtain mapping information. Clients may cache mapping information.

Resolver: A resolver assists a client to get the desired mapping information. A resolver knows one or more forest guides and may also employ a cache.

Tree: A tree is a hierarchy of LoST servers that hold the actual mapping data for a region (such as a country). The top of the tree is announced to one or more forest guides, and the leaves hold the authoritative data. Each tree is only responsible for one service for a particular

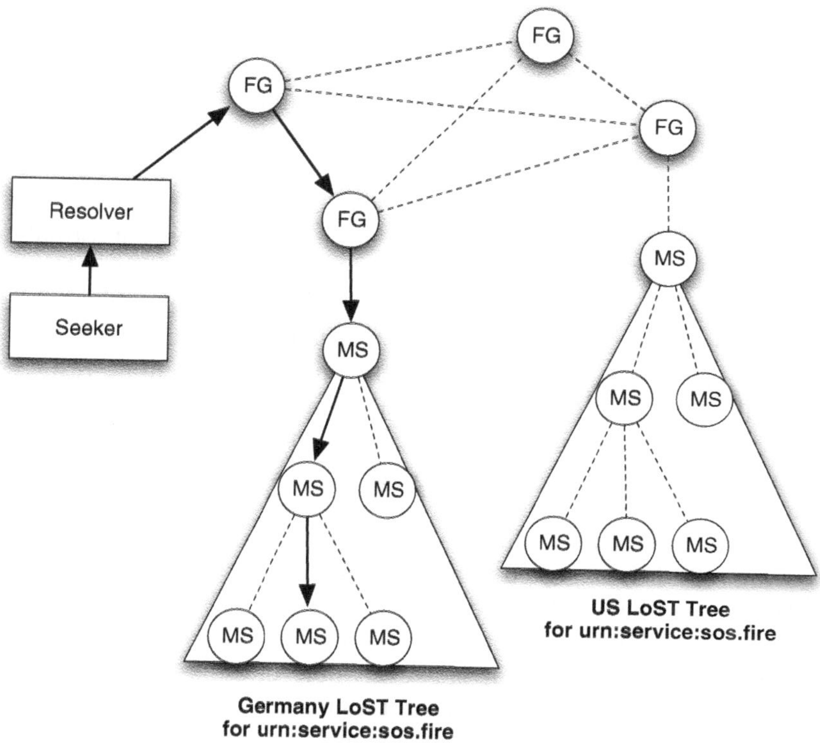

Figure 3.9 Overview of the LoST architecture, including Seekers, Resolvers, Forest Guides (FG), and Mapping Servers (MS).

region (e.g., a tree hosts mapping data for urn:service:sos.fire for Austria). Servers may participate in several trees if they have information about multiple services in a region, but from an architectural point of view, we consider the trees as separate things.

Forest guide: A forest guide maintains knowledge of the trees in the infrastructure and their coverage areas. To find out which tree has mapping data for a particular service and region, a forest guide has to be contacted. However, the forest guide does not contain actual mapping data.

A tree consists of authoritative mapping servers. These servers hold the actual mapping data and ultimately answer the findService requests. The tree may be organized hierarchically, so a non-authoritative mapping server may redirect or forward a particular request to another server, called a child, within the same tree. Note that there always has to be a single "root" server at the top of each tree (that means this mapping server has no parent but only child servers). The root is responsible for announcing the coverage region of this particular tree to forest guides. LoST servers can also be clustered for load balancing or availability. All LoST servers in a cluster share the same data set.

Another important part in the LoST architecture is the forest guide: They guide clients among the trees in the forest. The only information a forest guide has is a list of trees, holding their Service URNs and coverage areas (in civic and geodetic format, as available). The idea is that there will be a set of globally distributed forest guides, all having the same knowledge of the overall infrastructure and helping resolvers finding the right tree to perform a given mapping. In theory, it shouldn't matter which forest guide is contacted, the answer should be the same. However, some forest guides may apply a local policy resulting in different responses. This might be the case when, for example, there is a dispute over a territory and two trees claim to be authoritative for the same area. So it is expected that resolvers will query a preferred forest guide first, and have others configured for back-up purposes.

When the particular role of a LoST server isn't important, it is often referred to as a LoST server. However, note that the different servers have different responsibilities in the LoST architecture.

When looking at the LoST architecture, one might notice the similarity to the DNS. Indeed, there are some commonalities: Both the LoST infrastructure and the DNS have resolvers using hierarchies of servers to find authoritative information, with all participants speaking a common protocol. One difference is that the hierarchy is not based on labels (the domain name hierarchy in DNS), but rather on geographical coverage and service identifiers. Moreover, there is no single root for the LoST infrastructure as in DNS. This task is handled collectively by the forest guides. But there is not only one forest guide. In fact, anyone can operate a forest guide; one isn't committed to a single root. Although each forest guide should have the same set of data as every other forest guide (accomplished by LoST synchronization, described below), a forest guide may apply a local policy.

Each tree can be organized differently. The forest guide is only aware of the top of the tree and the coverage region of the complete tree. The tree itself consists of authoritative mapping servers. One typical deployment for emergency services could be as follows: there is one tree per country for each emergency service offered. For example, if the police is administered by the national government, there might be only one mapping server installed announcing that it is responsible for urn:service:sos.police for this country. Assuming that the fire brigade is organized by the federal states, one could deploy a LoST server for each federal state plus one LoST server at the top of the fire brigade tree. The mapping server at the top of the tree announces that it is responsible for urn:service:sos.fire for the whole country. However, this mapping server does not actually store mapping data but rather knows which child is responsible for which federal state. The mapping data is stored in the respective federal state servers. This example is shown in Figure 3.9. Federal state servers could then be further structured. For load balancing and availability, the mapping servers may be clustered. When a seeker located in Germany (or a resolver on behalf of the seeker) is looking for a mapping for urn:service:sos.fire in the scenario shown in Figure 3.9, (under the assumption that the mapping is not cached somewhere), the forest guide would redirect the seeker to the top of the tree, the server announcing it covers the tree for whole Germany. Then the request is sent there. Actually this server has no authoritative mapping information, but its child servers do. Hence, the mapping information is passed from the leaves of the tree up to the top. Finally, the seeker gets back mapping information from the tree via its local resolver.

3.4.7 Private and Public LoST Trees

LoST is a basic mapping protocol which is not limited to the usage scenarios described above. In fact, it may also be useful to utilize LoST for non-public data in the emergency services domain, for example, to map the location of the emergency caller to first responders. This information might not be exposed to the public via the public LoST infrastructure, but instead it could be stored in a separate tree that is only accessible from inside a private emergency services network. There have been some proposals to standardize Service URNs for such internal purposes, but since these LoST servers and services are operated within private networks, the need for interoperability is lower.

Using LoST for multiple purposes related to emergency calls may allow organizations that provide emergency services to benefit from some synergies. For example, PSAPs may use a single LoST client for multiple purposes, and because the LoST infrastructure has to be deployed anyway to allow the operation of the emergency calling framework as envisioned by the IETF, using it for additional purposes has a lower incremental cost.

3.4.8 LoST Synchronization

As described above, the LoST architecture assumes that there will be several sets of LoST servers deployed. For example, trees are organized hierarchically, and may be installed in clusters; and forest guides know about the coverage regions of trees. Consequently, the question arises of how these servers share information. The LoST protocol alone is not sufficient to synchronize LoST servers in a cluster or to exchange coverage regions among forest guides. These features are provided by the LoST synchronization mechanism (LoST-sync), again an XML-based protocol on top of HTTP, like the LoST base protocol, but used for different purposes. LoST-sync has three main areas of application:

- *Synchronizing mapping servers at the same hierarchical level in a cluster:* All the mapping servers in the cluster may synchronize their database using LoST-sync. Because such a cluster is a local matter, current proprietary database synchronization may suffice in some cases.
- *Synchronizing of forest guides:* Forest guides work with each other in order to get a complete view of all the trees in the forest. Exchanging coverage regions of trees is essential among forest guides to disseminate this information through the overall LoST architecture.
- *Pushing coverage regions up a tree:* Child mapping servers push the region they are responsible for up the tree to their parent. This step could also be done by manual configuration, especially for services where the coverage region of a service doesn't often change. However, in emergency fail-over scenarios, it is often desirable for LoST mappings to change quickly (to direct calls to a back-up PSAP), in which case an automated technique is preferable.

Note that LoST-sync allows two different types of information to be synchronized: coverage regions and complete mappings. Complete mappings are exchanged only when LoST-sync is utilized for synchronization of a LoST cluster. In the other two cases mentioned above, coverage regions are exchanged, not the complete mapping. The main difference between a mapping and a coverage region is that the coverage region does not contain contact information (URIs), dial strings and other additional information. When synchronizing coverage regions, only the Service URN, the service boundary as well as the responsible LoST server are given. The following examples will further clarify the capabilities of the LoST-sync protocol.

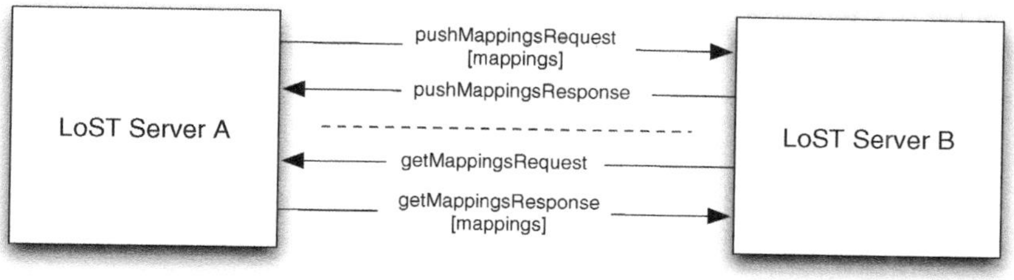

Figure 3.10 LoST synchronization messages.

Figure 3.10 shows a cluster consisting of two servers. The synchronization in this example is done via LoST-sync. When server A receives an updated mapping (e.g., via a new manual entry in its database), it notifies server B by pushing the mapping with a "pushMappingsRequest":

```
<?xml version="1.0" encoding="UTF-8"?>
<sync:pushMappingsRequest
  xmlns:sync="urn:ietf:params:xml:ns:lostsync1"
  xmlns="urn:ietf:params:xml:ns:lost1">
    <mapping source="lostserver.example"
      sourceId="7e3f40b098c711dbb6060800200c9a66"
      lastUpdated="2008-11-26T01:00:00Z"
      expires="2012-12-26T01:00:00Z">
      <displayName xml:lang="en">
        Example Police Department
      </displayName>
      <service>urn:service:sos.police</service>
      <serviceBoundary
  profile="urn:ietf:params:lost:location-profile:basic-civic">
        <civicAddress
  xmlns="urn:ietf:params:xml:ns:pidf:geopriv10:civicAddr">
          <country>EX</country>
          <A1>Example State</A1>
        </civicAddress>
      </serviceBoundary>
      <uri>sip:police@example.org</uri>
      <serviceNumber>112</serviceNumber>
    </mapping>
</sync:pushMappingsRequest>
```

Server B will store this mapping in its database as long as the pushed mapping is newer than its current mapping information. If the update process was successful, server B will send an answer:

```
<?xml version="1.0" encoding="UTF-8"?>
<pushMappingsResponse
  xmlns="urn:ietf:params:xml:ns:lostsync1"/>
```

However, just pushing new mappings around might not be sufficient. Imagine a new server is deployed or a server recovers after a breakdown. Hence, it is necessary to have a pull request in addition to the push request. The example shows such a pull request (called a “getMappingsRequest” in LoST-Sync) sent by a server that has an empty mappings database:

```
<?xml version="1.0" encoding="UTF-8"?>
<getMappingsRequest xmlns="urn:ietf:params:xml:ns:lostsync1"/>
```

In this way, a completely empty database could be populated with data sent in response by the peer. However, when there are already some mappings in the database, the LoST server should indicate which mappings it has with the “exists” element in the request, in order to avoid unnecessary data transfer:

```
<?xml version="1.0" encoding="UTF-8"?>
<getMappingsRequest xmlns="urn:ietf:params:xml:ns:lostsync1">
    <exists>
        <mapping-fingerprint source="lostserver.example"
        sourceId="08150815"
        lastUpdated="2010-10-10T09:09:09Z">
        </mapping-fingerprint>
    </exists>
</getMappingsRequest>
```

The response to this request will only contain updated mappings (newer than the time given in the “lastUpdated” attribute) of the existing one (identified by the “sourceId”) and any missing ones. A response is shown below:

```
<?xml version="1.0" encoding="UTF-8"?>
  xmlns:sync="urn:ietf:params:xml:ns:lostsync1"
  xmlns="urn:ietf:params:xml:ns:lost1">
  <mapping source="lostserver.example"
    sourceId="08150815"
    lastUpdated="2010-11-12T00:00:00Z"
    expires="2012-12-12T12:12:12">
      <!--actual mapping data omitted-->
  </mapping>
</sync:getMappingsResponse>
```

Care has to be taken when synchronizing peers: updates should be only accepted from trusted peers, so it's important to use a secure communication channel for synchronization. Furthermore, the receiver has to check whether it has already received this mapping information in order not to generate loops with circulating mappings. In all cases, incoming mappings should be checked if they do not violate any local policy or interfere with locally configured mappings.

Synchronizing forest guides works pretty much the same like synchronizing any other LoST mapping server. However, instead of complete mappings, forest guides just handle knowledge about where to find trees. So forest guides only synchronize coverage areas with their respective Service URN and the responsible LoST server for that tree. Hence, when using LoST-sync between forest guides or for pushing up a coverage region to a forest guide, the service boundary contains the coverage area of the tree. The URI element has to be empty, since the forest guide is not the entity performing the actual mapping (and besides, the actual URI is very likely to vary for locations inside this coverage area). Furthermore, the "serviceNumber" element is not needed at all. The LoST server responsible for the indicated coverage region is identified by the "source" attribute. So a forest guide is able to redirect an incoming request for this area and the given Service URN to the LoST server stated in the "source" attribute. Other servers down the tree may also be involved in answering the mapping request.

3.5 The Emergency Call Itself

As noted in the previous sections, it is necessary to perform several steps before an emergency call, in order to ensure that if an emergency happens, an emergency call can be placed without any trouble. In particular, the caller's device needs to have both location information and LoST mappings on hand.

The following sections discuss how an emergency call is actually placed, including the detection and routing of an actual emergency call as well as the structure of an emergency call SIP INVITE message.

3.5.1 Initiating Emergency Calls

The ECRIT architecture is not intended to change user behavior: As is usual today, emergency calls are initiated by dialing the emergency number known from the traditional phone system. Consequently, also VoIP phones have to detect by the dial string when the user is placing an emergency call (instead of a normal call). As was discussed in Section 3.3, Service URNs are generally not expected to be visible to the users, and of course at the moment nobody would type "urn:service:sos" to place an emergency call. On the other hand, Service URNs are used to route emergency calls, so the caller's device will need to know which dial strings correspond to which Service URNs. All the valid dial strings can be gathered with the help of the LoST protocol, from the "serviceNumber" element of the relevant LoST mapping. It is essential that the answer from the LoST server includes the dial string for an emergency service, even though the element was defined optional by the IETF. Without knowledge of the emergency dial strings it is impossible for a calling device to detect an emergency call. Since there are many different emergency dial strings throughout the world, it might be hard for a user to remember the local emergency dial string for the current location when traveling. To assist a user in such a case, so-called "home emergency dial strings" can be configured into a device. These dial strings are those from the home country of the user, and thus probably the ones the user is best able to remember even in a stressful emergency situation. A device configured with home emergency dial strings in addition to the local emergency dial strings, would allow a user to initiate an emergency call with either set. For example, a device sold in Austria might have 133 as home emergency dial string for the police. During a visit abroad in Germany, the user could call the local police by dialing either 110 (the local emergency dial string for the police in Germany) or 133 (the home emergency dial string).

The home emergency dial string is an optional way to initiate an emergency call; the local emergency number is always required to work. The home emergency dial string can also be configured via LoST, if a user has configured the device with his home country – the device can simply use a LoST query that contains the country code as location information. The normal process of listing all available emergency services in the home country, followed by a mapping query for each of them, can be used for home numbers as well as local numbers (for more details, refer to Section 3.6). However, it has to be noted that an emergency dial string of one country might be a regular phone number in another country, which could lead the user to accidentally place an emergency call by trying to place a local call. Since the user is familiar with his home emergency dial string, he might not carelessly dial this number, but it might nonetheless be necessary to temporarily override the home emergency dial string (just for one call), so that the caller can reach the local regular phone number.

The re-use of the PSTN notion of dial strings is deliberate: It matches a user interface paradigm that users are familiar with (namely, dialing a well-known number), and it helps prevent the user from making emergency calls by accident. Eventually, there may be devices that can place emergency calls, but do not have a traditional dialing interface; some of the ECRIT calling clients in Chapter 6 have this property. These devices might implement an emergency button, but should employ steps to reduce the risk of unintended calls. For example, after the emergency button is pressed, the device might prompt the user to confirm that an emergency call is indeed intended.

Right after an emergency call has been detected, a connection to the PSAP should be established as soon as possible so that the caller does not panic. Updating location and mapping information right before the actual emergency call is feasible when it can be done quickly, but it is generally preferable for the calling device to periodically refresh this information so that it has reasonably up-to-date information already on hand when an emergency call is placed.

When an emergency call is placed, the phone has to take certain precautions, so that the emergency call and a possible callback from the PSAP can be conducted without problems. A few important examples include:

1. Deactivation of call forwarding and do-not-disturb functions (to allow for possible callbacks from the PSAP).
2. Prevention of unintended hang-ups by the caller (e.g., by deactivation of the hang-up button on the user interface) or alerting the user to pick up the handset again. The reason for this is that the emergency calls should be ended by the PSAP call taker and not by the caller.
3. Deactivation of call hold features (so that the user does not put an emergency call on hold by mistake)

A full list of requirements is given in the IETF documents that describe the overall ECRIT requirements (Rosen and Polk, 2010), and its more narrative companion document contains some guidance and explanations (Rosen et al., 2010). Depending on how VoIP services are deployed, some of these requirements can affect both the VoIP provider and the user's phone at the same time. For example, call forwarding may be configured at the VoIP provider as well as on the phone, and whichever of these entities supports the feature has to disable it for an emergency call.

The ECRIT architecture also defines a way to test emergency calls (a functionality that does not exist at all in the traditional phone system). Using this mechanism, it is possible for the user or the calling device to automatically check that the emergency call functionality works. In order to indicate a test, the Service URN is extended by ".test"; everything else stays the same. The PSAP can detect that an incoming call is a test call by looking for this extension to the Service URN, and can use this information to reply with an automated response service rather than a call taker. So it is possible for users or devices to check whether the communication to the PSAP would be possible, in advance of an emergency call.

3.5.2 Routing Emergency Calls

Once a device has recognized that an emergency call has been placed, it needs to route the call. There are two ways that a call can get routed to a PSAP:

1. Via the VoIP provider's infrastructure.
2. Directly (without the indirect route via a VoIP provider).

Possibility 1 can be only considered when the user's phone has been configured with the information necessary to use a VoIP provider (including correct username and password). Using a VoIP provider can increase the likelihood of a call succeeding. For example, since normal calls are also passed to the VoIP provider, it is very likely that also the establishment of a connection is not being blocked by firewall rules. However, the VoIP provider needs to be able to handle emergency calls, for example, to have proxies that can correctly handle SIP messages with multi-part bodies. In some cases, it would even be imaginable that the VoIP provider could send additional information about the user to the PSAP, like the home address of the user. Such additional information has not yet been standardized, but some jurisdictions already require a VoIP provider to add a P-Asserted-Identity (Jennings et al., 2002) and/or an Identity header (Peterson and Jennings, 2006) in order to indicate the identity of the user.

If one does not want to depend on having a VoIP provider, the connection to the PSAP can be established directly (possibility 2 above). After collecting location information and LoST mappings, the phone itself has all the necessary information to send an emergency call directly to the PSAP. So the phone could call the PSAP directly, bypassing the VoIP provider. In this case, even if the caller normally uses a VoIP provider, it doesn't matter whether the VoIP provider supports the ECRIT emergency call framework or not, since the emergency call wouldn't be processed by any of the provider's servers. However, the burden then is on the calling device to make sure that a firewall or NAT device does not block this connection attempt. Possibility 2 also works if a user has no VoIP provider at all, or in the case of configuration errors, including lost passwords.

Finally, it should be noted that SIP proxy servers are supposed to apply the normal routing decisions as defined for SIP (Rosenberg and Schulzrinne, 2002) for emergency calls as well as normal calls.

3.5.3 Assembling the SIP INVITE Message

A SIP emergency call is set up with a SIP INVITE message that contains the following information:

- Service URN.
- SIP URI of the PSAP.
- SIP URI of the caller.
- Location information (either by Value or by Reference).

The ECRIT architecture document (Rosen and Polk, 2010) explains in detail how the signaling of SIP-based emergency calls has to be achieved. The most important requirement is that certain fields in the header of the SIP INVITE for the call contain specific values:

- Request URI: Service URN for the call.
- To header: Service URN for the call.
- From header: SIP "Address of Record" for the caller.
- Route header: SIP URI for the PSAP.
- Contact header: Current SIP contact information of the client (to support callbacks).
- Supported header: Must contain the value "geolocation".
- Geolocation header: Location information with any necessary parameters, if location information is available.

The header of a SIP INVITE message following these rules is shown below (note that some headers are omitted for clarity):

```
INVITE urn:service:sos.police SIP/2.0
Route: <sip:police@example.at;lr>
Via: SIP/2.0/TCP 192.0.2.1:5060;rport;branch=z9hG4bK4r1id0
To: "Vienna Police" <urn:service:sos.police>
From: John Doe <sip:jd@example.com>;tag=1nexm5g
CSeq: 2 INVITE
Contact: John Doe <sip:jd@192.0.2.1;grid=74500e587>
Supported: geolocation
Geolocation: <cid:pidflo@zap>;
  inserted-by=192.0.2.1;recipient=endpoint;used-for-routing
```

The body of the SIP message is not shown, but is expected to contain SDP information that describes the multimedia streams that will make up the call, as well as a PIDF-LO location object (indicated by the Geolocation header). The SDP information is generated in the same way as for normal calls, and the way location information is embedded in the message is described in Section 3.2.3.

3.6 Home Dial String Configuration via LoST

Home dial strings can be convenient for the users, but it can be prone to errors and frustrating to configure them manually. Imagine, a user has to first select the kind of emergency services available in his home country.

Second, all the (hopefully) correct dial strings have to be entered. For countries where multiple dial strings are in use, it is more than likely that users won't successfully perform this task. Hence, it is much more convenient to just let the user select his home country and have his calling device perform the procedure below to figure out dial strings automatically via LoST.

The idea is to use the country code of the home country of the user as location information input for the LoST queries. (This location information might not be sufficiently precise in all cases, but works in most countries around the world.) The actual mapping response can be disregarded except for the "serviceNumber", which contains the home emergency dial string.

The first step is a query for a list of all the emergency services in the home country (i.e., the sub-services of the "urn:service:sos" Service URN):


```
<?xml version="1.0" encoding="UTF-8"?>
<listServicesByLocation
  xmlns="urn:ietf:params:xml:ns:lost1"
  recursive="true">
  <location id="homecountry" profile="civic">
    <civicAddress
      xmlns="urn:ietf:params:xml:ns:pidf:geopriv10:civicAddr">
      <country>AT</country>
    </civicAddress>
  </location>
  <service>urn:service:sos</service>
</listServicesByLocation>
```


In this example, the home country is set to Austria (AT). The server answers with a list of services:


```
<?xml version="1.0" encoding="UTF-8"?>
<listServicesByLocationResponse
 xmlns="urn:ietf:params:xml:ns:lost1">
 <serviceList>
  urn:service:sos.ambulance
  urn:service:sos.fire
  urn:service:sos.police
 </serviceList>
 <path>
  <via source="resolver.example"/>
```

```
    <via source="authoritative.example"/>
  </path>
  <locationUsed id="homecountry"/>
</listServicesByLocationResponse>
```

In this example, ambulance, fire brigade, and police services are available (actually there are more emergency services in Austria which are not shown in this example). Hence, the device needs to obtain a mapping for each of these services, as well as for "urn:service:sos" itself (since top-level services aren't listed in the response!). The example below shows the mapping request for the police:

```
<?xml version="1.0" encoding="UTF-8"?>
<findService xmlns="urn:ietf:params:xml:ns:lost1"
  recursive="true" serviceBoundary="reference">
  <location id="homecountry" profile="civic">
    <civicAddress
      xmlns="urn:ietf:params:xml:ns:pidf:geopriv10:civicAddr">
      <country>AT</country>
    </civicAddress>
  </location>
  <service>urn:service:sos.police</service>
</findService>
```

Again, the country code "AT" is used as location information. To reduce unnecessary traffic, the service boundary should be requested by reference only (since the boundary information isn't of interest at all). The response from the LoST server may contain a warning since the location information is not precise enough. This warning can usually be ignored, though, since the actual mapping isn't of interest anyway, just the service number of the response is important:

```
<?xml version="1.0" encoding="UTF-8"?>
<findServiceResponse xmlns="urn:ietf:params:xml:ns:lost1"
  <mapping
    expires="2011-01-01T01:44:33Z"
    lastUpdated="2009-11-01T01:00:00Z"
    source="authoritative.example"
    sourceId="7e3f40b098c711dbb6060800200c9a66">
```

```
      <displayName xml:lang="en">
        Police Vienna - Default PSAP for Austria
      </displayName>
      <service>urn:service:sos.police</service>
      <serviceBoundaryReference
        source="authoritative.example"
        key="7214148E0433AFE2FA2D48003D31172E"/>
      <uri>sip:police@example.at</uri>
      <serviceNumber>133</serviceNumber>
    </mapping>
    <path>
      <via source="resolver.example"/>
      <via source="authoritative.example"/>
    </path>
    <locationUsed id="homecountry"/>
  </findServiceResponse>
```

The client just has to extract the "serviceNumber" (in this case, 133) and store a connection between that dial string and the Service URN (in this case, "urn:service:sos.police"). All the other information must be dropped, since it won't be relevant to the user's current location. After the mapping procedure is repeated for each service, to gather all the home emergency dial strings, the client is able to detect an emergency call when the user dials his well-known home emergency dial strings.

In particular, the client device will be able to detect which emergency service the user is requesting when he dials a home dial string, since it can look up the corresponding URN. Once the client device has made this mapping, though, it must send the call to the *local* PSAP URI for that URN, this is, the PSAP URI that it obtained using current location information – not the PSAP URI obtained in the mapping request for the home dial string. If the requested emergency service doesn't have a local equivalent, then the device should send the call to the local PSAP for the general emergency service "urn:service:sos".

The procedure described above works automatically; only the home country of the user has to be configured. Certainly, configuring only the home country is much easier for a user than manually configuring all of the different dial strings.

Finally, keep in mind that in most jurisdictions, the home emergency dial string is optional, while the local dial string is mandatory.

3.7 Deployment Models

The ECRIT architecture specifies three basic steps in an emergency call:

1. Locating the caller.
2. Location-to-service mapping.
3. SIP call routing to the PSAP.

Which entities take responsibility for these tasks is not prescribed by the standards. Rather, the basic protocols used to accomplish these steps – location configuration protocols, LoST, and SIP – can be used in a few different configurations to accomplish an emergency call.

The basic question is whether the caller's device or an element of the VoIP network (i.e., a proxy) performs the geolocation and mapping functions. Since geolocation must precede mapping, there are three possible scenarios:

1. Calling device does both geolocation and mapping.
2. Proxy does both geolocation and mapping.
3. Calling device does geolocation; proxy does mapping.

In protocol terms, the main difference between these scenarios is which entities interact with the location and mapping servers (and thus how these servers have to be discovered), and how the fields in the SIP INVITE sent from the caller to the VoIP provider are populated. (The message sent from the provider to the PSAP should always have the form specified above, with the Request URI, Geolocation, and Route headers properly populated.) Table 3.4 provides a summary. Let's look at each scenario in a little more detail, then consider what the differences mean for devices and VoIP providers.

The situation where the calling device performs both geolocation and mapping functions is the one we mostly focus on this book, and indeed the scenario that most people envision when they think about ECRIT emergency calling. This scenario is in some ways the

Table 3.4 A taxonomy of ECRIT deployment scenarios

Location	Mapping	Request-URI	Geolocation	Notional Network Type
Caller	Caller	Service URN	Provided	Pure VoIP
Proxy	Proxy	Dial string	Empty	PSTN-like, enterprise
Caller	Proxy	Dial string	Provided	

simplest. All of the ECRIT-specific things that need to be implemented (i.e., location and LoST servers) need only be accessible locally, to callers and PSAPs. The device has all the information it needs, not only to place an emergency call, but even to recognize that one has been dialed. The major drawback of this scenario is that it requires endpoints to be very capable, in that they will have to have software to perform location and mapping functions, and enough bandwidth and power resources to keep this information up to date. While these are not very onerous requirements, they do require that endpoints that are not already ECRIT-compliant be upgraded. On the other hand, there is no need to upgrade elements in the calling network in this scenario, unless they fail to support some parts of the SIP specification required for emergency calling, for example, multi-part SIP message bodies or Service URNs as Request URIs. Because there is no requirement in this case for the core VoIP network to have access to location or mapping servers, this solution is likely to work well for "over-the-top" or "pure VoIP" systems (like Skype or Vonage) whose caller access systems aren't bound to specific physical networks. These networks also tend to have a greater ability to upgrade clients, for example, as soft-phones on a general PC platform.

Handing all the labor to the proxy provides the opposite set of trade-offs: Changes will be needed in the calling network (to support additional call routing functions), but edge devices can remain the same. In this system, all the calling device does is collect dialed digits from the user and pass them to the network in a SIP INVITE. The network receives the call, recognizes the dial string, and routes it to the correct PSAP. This system creates a few technical issues. The first question is when the network should invoke special emergency call handling, that is, how it should recognize emergency calls. Since emergency numbers depend on the caller's location, a general solution of this form would have to perform location and LoST queries for every call, in order to get local dial strings and check them against the dial string in the INVITE. The second question is how the VoIP network will get access to information about the caller's geolocation. If the VoIP network is tied into a specific physical network, then it may be able to have obtain this information directly from the physical network infrastructure, but otherwise, the VoIP provider will need a means to discover a Location Server that can provide information on the caller (e.g., a server for the caller's current access network).

A pure network-based solution is thus best adapted for "vertically integrated" networks that integrate a VoIP calling function with physical- and network-layer access. This integration both forces a geographical constraint on where clients can be (so the network can cache emergency dial strings) and provides the VoIP network with a way to locate clients. Edge calling devices on these networks also tend to be more difficult to upgrade with new functionality. Several current VoIP networks have these properties, including many PSTN-replacement services and enterprise VoIP systems; indeed, many of these networks already have geolocation and mapping systems that can be adapted to take advantage of new LoST databases and send ECRIT-compliant messages to PSAPs.

The combination of terminal-provided location with network-side mapping is an intermediate case that isn't as clearly advantageous for any general class of networks. Because emergency call recognition and mapping are still done in the network, the same issues with call recognition arise as above. The main difference between this case and the one where the network also does geolocation is that the network does not have to figure out how to get geolocation information, and thus can more easily handle a client base that can access the VoIP service from many networks. As one particular example, the emergency calling architecture for the 3GPP IP Multimedia Subsystem (*IP Multimedia Subsystem (IMS) Emergency Sessions*, 2010) (the VoIP standard for next-generation cellular networks) uses a combination of the latter two cases. Calls that are recognized as emergency calls (via an unspecified mechanism, likely a fixed list of numbers) are routed to an "Emergency Call Session Control Function" (E-CSCF), which is an SIP proxy dedicated to handling VoIP calls; the E-CSCF obtains routing information from a "Location Retrieval Function". The function of the LRF is currently unspecified, but in at least one current implementation (the Fraunhofer FOKUS Open IMS Core), the LRF obtains routing information from the LoST infrastructure.

The fact that these three different scenarios exist creates an interoperability challenge. If a dumb terminal that does not support any emergency calling functions is connected to a VoIP provider that requires endpoints to perform both geolocation and mapping, then its emergency calls will not go through. And if an endpoint that performs geolocation and mapping connects to a VoIP provider that does not allow calls with Service URNs in the request URI, then its calls will also fail.

The best way to avoid these failures is for devices and networks to follow the "Postel robustness principle": Be conservative in what you do, and be liberal in what you accept from others. When possible, endpoints should implement many of the above behaviors, and use the ECRIT test calling function discussed in Section 3.5 to determine what types of calls are supported. VoIP providers should keep in mind that multi-network clients may attach to their networks – imagine a mobile device that uses a 3GPP IMS network on its cellular interface and pure VoIP service when using WiFi or WiMAX. Given this fact, VoIP networks should support as many as the above use cases as possible, and provide clear error indications for cases they do not support.

3.8 Considerations for Proxies

The ECRIT architecture also allows entities in the VoIP provider's network to perform call routing on behalf of endpoints, including recognizing that an emergency call has been placed, looking up the endpoint's location and using LoST to find the proper PSAP. Proxies are required to perform routing in some new architectures (in particular, the 3GPP IMS architecture for emergency calling (*IP Multimedia Subsystem (IMS) Emergency Sessions*, 2010)), but even outside these architectures, there will still be a need for routing proxies to support end devices that are not yet ECRIT-enabled. The security considerations for a proxy that performs emergency call routing are largely the same as when the endpoint performs these functions, but there are some important differences.

The most basic challenge for a routing proxy is how to distinguish emergency calls from non-emergency calls, if all the endpoint provides is a set of dialed digits. Endpoints can discover their local emergency numbers using LoST, but when a proxy receives a call from a non-ECRIT endpoint, it will not have location information attached. So the only approach for proxies that addresses the general case is for the proxy to look up location information for every call, then look up the corresponding emergency numbers and compare against the dialed digits. In certain special situations, there are intermediate solutions. For instance, if a proxy can only be accessed by callers in a given geographical area, then it can be statically configured with emergency numbers from that area. (However, even then, care needs to be taken to ensure that the assumption of geographical limitations is actually valid!)

Determining the caller's location is another challenge, outside of a few special cases. The general problem is that the proxy must find a Location Server that has information about the caller, and then it must be authorized to ask for that information. In the special circumstance where the same entity controls both the VoIP network and the underlying IP network (as is common in many "PSTN-replacement" services today), the proxy can be statically configured to access the Location Server for that network. In the slightly more general case where a proxy covers a prescribed geographic area, it may be possible for the proxy to have pre-configured trust relationships with Location Servers covering that area.

In the general case where a proxy accepts calls from anywhere on the Internet, such a proxy may sometimes be able to discover a Location Server for the caller using the reverse DNS. One way that networks can advertise Location Servers is by putting LIS discovery records (as described in Section 4.1) in the reverse-DNS tree. End devices can use these records to look up their own Location Servers, but third parties like proxies can as well. Assuming the network operator has published these records, they will allow a proxy to look up a Location Server based on the caller's public IP address. This solution still leaves two open questions, namely (1) whether the public IP address that the proxy sees actually identifies the caller (as opposed to a proxy or NAT), and (2) whether the proxy is authorized to query the Location Server.

Of course, if the caller's local access network is relying on either DHCP or layer-2 mechanisms to deliver location for use in emergency calling, then a proxy in some other network has no hope of being able to access this information; in such networks, the calling device must at least provide location information.

For LoST mapping, in principle, there should be no difference between the endpoint performing a LoST lookup and a proxy performing a LoST lookup, however, in practice, there is likely to be a difference. The operative principle here is that the LoST infrastructure will constitute a single, globally consistent database like the DNS, in particular, one that does not return different results to different requesters. Especially during the initial deployment phases of the LoST infrastructure, however, it is likely that different requesters will get very different views of the LoST database, with some getting accurate results and some getting no results at all. For example, if LoST is deployed as a local service, an end device may be able to discover its local LoST server over DHCP and ask it for mappings without any problems, but a proxy might not even be able to figure out what server to ask.

This incongruity leads to two important observations: First, in the short run, it will be more reliable for LoST mapping to be done by endpoints. Second, in the longer run, as the LoST infrastructure is deployed, organizations that operate LoST servers should work together to make their servers discoverable to one another, according to the LoST mapping architecture (Schulzrinne, 2009). There are a couple of paradoxes here – endpoints that can do LoST mapping make global consistency unnecessary, but routing proxies (which require global consistency) are used precisely because endpoints cannot do LoST mapping – but these reflect fundamental limitations of the routing proxy model.

3.9 Standardization

The IETF is the standards organization for the Internet. An overwhelming majority of protocols in use in the Internet today were standardized by the IETF. Besides the IETF, several other organizations are working on emergency calls, including for example, ETSI EMTEL, IEEE, Wimax Forum, Open Mobile Alliance, ITU-T, OASIS, 3GPP, NENA and the Broadband Forum. Most of these organizations use the IETF ECRIT emergency calling framework at the IP layer, and focus on related standards at lower layers (e.g., in the Broadband Forum and IEEE) or higher layers (e.g., in NENA and ETSI EMTEL). Some organizations have also worked on adding detail to the ECRIT specifications, for example, specifying the order in which location protocols should be tried.

Within the IETF, the primary working groups related to emergency calling are GEOPRIV and ECRIT. ECRIT is chartered with defining an architecture for emergency calling, while GEOPRIV focuses on geolocation and privacy more generally. The SIP working group (now called SIPCORE) developed the mechanism for location conveyance in SIP.

The status of the working groups is best viewed at:

- `http://tools.ietf.org/wg/ecrit/`.
- `http://tools.ietf.org/wg/geopriv/`.
- `http://tools.ietf.org/wg/sipcore/`.

The most important documents in GEOPRIV and ECRIT are the following:

- draft-ietf-ecrit-framework;
- draft-ietf-ecrit-phonebcp;
- RFC 5582 (Location-to-URL Mapping Architecture and Framework);
- RFC 5222 (LoST);

- RFC 5223 (Discovering LoST Servers using DHCP);
- RFC 5031 (Service URNs);
- RFC 5985 (HTTP-Enabled Location Delivery (HELD));
- RFC 5986 (Discovering local Location Servers);
- RFC 4119, RFC 5139, RFC 5491 (PIDF-LO);
- RFC 5774 (Considerations for Civic Addresses);
- RFC 3825, RFC 4776 (DHCP location configuration).

In the jargon of the IETF, RFCs are permanent, archival documents that reflect the consensus of the IETF, while Internet-drafts are working documents (which often eventually become RFCs); all RFCs start as Internet-drafts. All of these documents can be quickly accessed by name using the IETF document retrieval service. The document with name `document-name` can be accessed at the URI `http://tools.ietf.org/html/document-name`. For example:

- draft-ietf-ecrit-framework can be found at `http://tools.ietf.org/html/draft-ietf-ecrit-framework`.
- RFC 5582 can be found at `http://tools.ietf.org/html/rfc5582`.

Currently, ECRIT and GEOPRIV are still refining these standards and developing new documents to address a few remaining use cases. There are thus more documents being developed than are listed above, so the reader should refer to the working group status site to get a complete overview. The ECRIT "big-picture" documents (Rosen et al., 2010; Rosen and Polk, 2010) can be used as an entrance point.

The most important document in the SIPCORE working group for emergency calls is of course the document that defines location conveyance using SIP (Polk and Rosen, 2010).

Standards at the link layer (layer 2) can be important for emergency calling because layer-2 entities can sometimes provide high-quality geolocation information to endpoints. The ECRIT framework documents use LLDP-MED – the combination of an IEEE Link Layer Discovery Protocol with extensions for Media Endpoint Discovery defined by TIA – for the purpose of location configuration. LLDP-MED itself makes use of the IETF standards to describe location information (drawing on the standards to convey location in DHCP). Some IEEE working groups are working on further improvements for emergency calls (e.g., support to allow unauthenticated users access to the layer-2 network for emergency calls).

The Broadband Forum is a worldwide consortium of companies that provide broadband telecommunications service and equipment. (The Broadband Forum was formerly known as the DSL forum, and is widely known for developing international standards for DSL, ADSL, VDSL, etc.) The Broadband Forum has a technical working group on the topic of emergency calls via VoIP. This group has built on the ECRIT architecture by adding a layer of detail. For instance, the requirements they have defined for Customer Premises Equipment (CPE) have much more detail than in the corresponding IETF requirements document (Rosen and Polk, 2010). In addition to requirements, they have even concrete sequences of operations, as described in the Broadband Forum document "Requirements for CPE in Support of Accessing Emergency Services" (*Requirements for CPE in Support of Accessing Emergency Services (work in progress)*, 2010).

The National Emergency Number Association (NENA) has long been the leading body for coordination among PSAPs and emergency services vendors in the US. NENA supports the development of the emergency calling systems in North America and has been working on forming the next generation of emergency call systems. Various NENA standards and documents are available at `http://www.nena.org/`. The most important NENA standard for VoIP emergency calling is the "i3" architecture, which defines an all-IP emergency calling system based on the ECRIT architecture. They have also defined an intermediate architectures known as "i2". NENA also provides an informational website for emergency calls over VoIP, `http://www.911voip.org/`. The European Emergency Number Association (EENA), the European equivalent to NENA, which has traditionally focused on the development of the single European emergency number 112, recently formed a technical committee focused on coordinating VoIP emergency calling solutions through Europe.

3.10 Summary

The ECRIT architecture brings together many different technologies to accomplish its three basic steps of locating the caller, finding the right PSAP, and placing an emergency call. Figure 3.11 is a complete reference illustration of where each piece fits in the overall architecture.

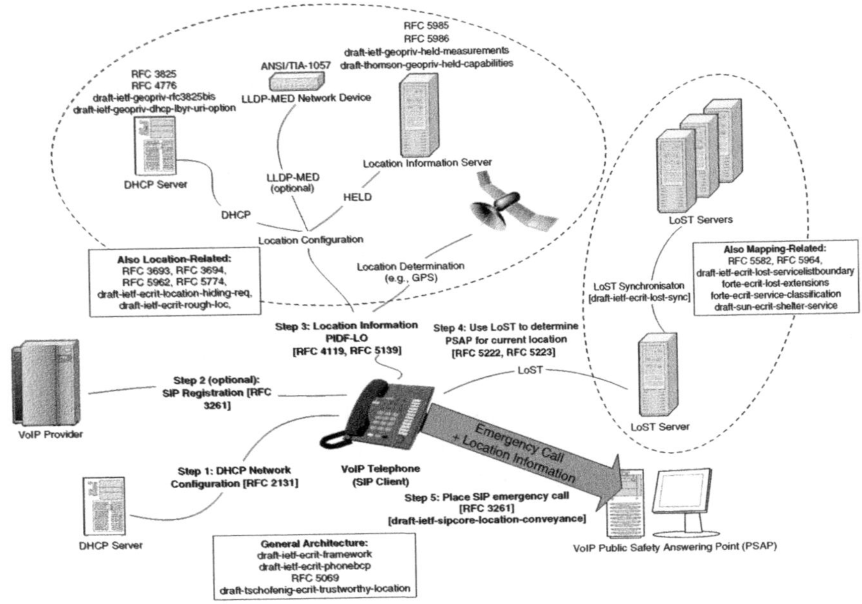

Figure 3.11 Summary of the ECRIT architecture.

References

Handley M, Jacobson V and Perkins C (2006) SDP: Session Description Protocol. RFC 4566.

Hardie T, Newton A, Schulzrinne H and Tschofenig H (2008) LoST: A Location-to-Service Translation Protocol. RFC 5222.

IP Multimedia Subsystem (IMS) Emergency Sessions (2010). Technical report, 3GPP.

ISO 8601:2000. Data Elements and Interchange Formats – Information Interchange – Representation of Dates and Times (2000). ISO.

Jennings C, Peterson J and Watson M (2002) Private Extensions to the Session Initiation Protocol (SIP) for Asserted Identity within Trusted Networks. RFC 3325.

Klyne G and Newman C (2002) Date and Time on the Internet: Timestamps. RFC 3339.

Levinson E (1998) Content-ID and Message-ID Uniform Resource Locators. RFC 2392.

Masinter L (1998) The "data" URL scheme. RFC 2397.

Peterson J (2005) A Presence-based GEOPRIV Location Object Format. RFC 4119.

Peterson J and Jennings C (2006) Enhancements for Authenticated Identity Management in the Session Initiation Protocol (SIP). RFC 4474.

Polk J and Rosen B (2010) Location Conveyance for the Session Initiation Protocol. Internet Draft (work in progress) draft-ietf-sipcore-location-conveyance.

Requirements for CPE in Support of Accessing Emergency Services (work in progress) (2010). Technical report, Broadband Forum.

Rosen B and Polk J (2010) Best Current Practice for Communications Services in Support of Emergency Calling. Internet Draft (work in progress) draft-ietf-ecrit-phonebcp.

Rosen B, Schulzrinne H, Polk J and Newton A (2010) Framework for Emergency Calling Using Internet Multimedia. Internet Draft (work in progress) draft-ietf-ecrit-framework.

Rosenberg J and Schulzrinne H (2002) Session Initiation Protocol (SIP): Locating SIP Servers. RFC 3263.

Schulzrinne H (2006) Dynamic Host Configuration Protocol (DHCPv4 and DHCPv6) Option for Civic Addresses Configuration Information. RFC 4776.

Schulzrinne H (2008) A Uniform Resource Name (URN) for Emergency and Other Well-Known Services. RFC 5031.

Schulzrinne H (2009) Location-to-URL Mapping Architecture and Framework. RFC 5582.

Schulzrinne H, Polk J and Tschofenig H (2008) Discovering Location-to-Service Translation (LoST) Servers Using the Dynamic Host Configuration Protocol (DHCP). RFC 5223.

Sugano H, Fujimoto S, Klyne G, Bateman A, Carr W and Peterson J (2004) Presence Information Data Format (PIDF). RFC 3863.

Thomson M and Winterbottom J (2008) Revised Civic Location Format for Presence Information Data Format Location Object (PIDF-LO). RFC 5139.

Winterbottom J, Thomson M and Tschofenig H (2009) GEOPRIV Presence Information Data Format Location Object (PIDF-LO) Usage Clarification, Considerations, and Recommendations. RFC 5491.

Wolf K and Mayrhofer A (2010) Considerations for Civic Addresses in the Presence Information Data Format Location Object (PIDF-LO): Guidelines and IANA Registry Definition. RFC 5774.

4

Including Location Information

Location information plays a central role in the ECRIT emergency calling architecture. An emergency call over the Internet, according to the ECRIT architecture, is only possible if the telephone originating the call (or some entity acting on its behalf, such as a SIP proxy) knows the caller's location. This chapter is therefore devoted to this important topic.

The first distinction to make here is between two critical operations, location determination and location configuration. *Location determination* is the process of actually determining or measuring the location of an endpoint, for example, via a GPS unit. The result of location determination is location information, encoded in a location object. The process of *location configuration* "configures" location information into a device – either manually through user entry or automatically (e.g., with the help of a Location Server). In cases where a Location Server is used, the server itself must have access to some location determination mechanism (e.g., via triangulation in a wireless network or from a database for a fixed network), so that it can provide location information to configure the client endpoints that query it. The complexity of location determination is the price one pays for the mobility of VoIP subscribers.

In addition to emergency calls, location information can also be used for other interesting applications. The protocols and concepts described in this chapter are not limited to emergency calling applications (or even other types of emergency or calling applications); rather, they can be used to support other location-based services in exactly the same way as emergency services.

VoIP Emergency Calling: Foundations and Practice Karl Heinz Wolf and Richard Barnes

4.1 Location Configuration

Location configuration happens when location information is either entered into a device by the user himself, or made available to the device through some Location Configuration Protocol (LCP). The IETF has defined two protocols for automatic location configuration: They have extended the well-known Dynamic Host Configuration Protocol (DHCP) and developed a specific location protocol called HTTP-Enabled Location Delivery (HELD). In addition, the ECRIT emergency calling architecture supports the IEEE/TIA LLDP-MED protocol. We will provide an overview of these three protocols in the following sections. Note that these protocols are independent of determination; how location information is actually found is beyond the scope of this section.

Naturally, users can also enter their location themselves to manually configure their telephone (assuming that the telephone's user interface supports it) – with manual configuration, however, there is always a risk that location is never configured, or configured incorrectly. For example, a user might enter a location when he first sets up a device, but not when he moves it later. Having false location information can cause several problems for an emergency call, delaying or even preventing the delivery of the required aid. Thus, we will focus on automatic location configuration for the remainder of this section.

4.1.1 HTTP Enabled Location Delivery (HELD)

HELD is an application-layer (layer-7) protocol for location configuration, which uses HTTP or HTTPS as a transport (see the protocol stack in Figure 4.1). HELD was specified by the IETF as an XML-based

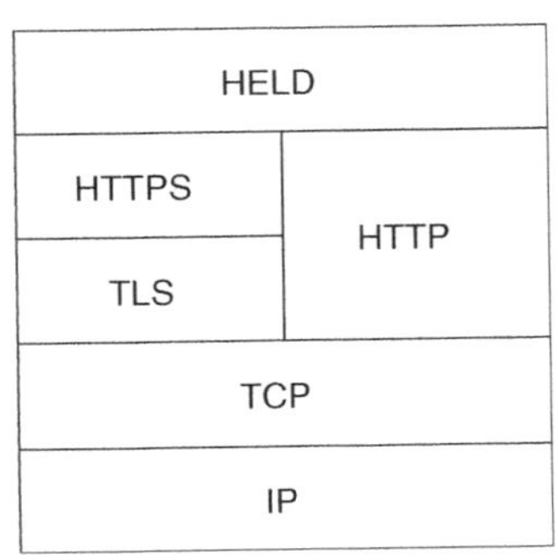

Figure 4.1 The HELD protocol stack.

protocol for a client to communicate with a Location Server. In contrast to some other proposed protocols, HELD was developed specifically for location configuration (as opposed to, say, third-party requests for location).

One primary use case envisioned in the IETF location system is the case in which a Location Server operated by a network operator (i.e., an ISP) would provide a location configuration service to that network's subscribers. Location information can be provided either "by value" or "by reference" (see also Section 3.2.2). When the Location Server delivers location by value, it sends the client a PIDF-LO document (inside of a "locationResponse" element), while location by reference is delivered as a URI that points to a PIDF-LO document. If one wants to take location information acquired via HELD and transmit it using SIP (see Section 3.2.3), there is no conversion necessary, since the PIDF-LO can be copied directly into the body of a SIP message.

A Location Server must be able to identify the clients in its network in order to deliver the correct location information to each device. The simplest possibility is to identify a client based on its IP address (which the server can see as the source IP address of a HELD request). Of course, this approach is not without problems; one thinks, for example, of Network Address Translation (NAT), where several clients are found behind one IP address. In the common scenario where a NAT device is incorporated in the network gateway for a home, the Location Server will not be able to determine location more precisely than the address of the house; however, more precise information is generally not required for routing emergency calls. So there are some cases where combining the use of IP addresses with NAT devices doesn't cause any problems. The use of Virtual Private Networks (VPNs) is a larger problem. When a device uses a VPN (e.g., to connect to a corporate network), it must perform location configuration before establishing the VPN connection. This is because VPNs frequently connect endpoints to physically distant locations, which means that performing location configuration *through* the VPN would frequently result in the client receiving incorrect location information. The IETF is currently discussing the use of identifiers other than IP addresses.

To illustrate how HELD works, here is an example of a simple HELD location request:

```
POST /location HTTP/1.1
Host: lis.example.com
Accept: application/held+xml,
```

```
  application/xml;q=0.8,
  text/xml;q=0.7
Content-Type: application/held+xml

<?xml version="1.0"?>
<locationRequest
         xmlns="urn:ietf:params:xml:ns:geopriv:held"/>
```

This simple HTTP POST request shows how the client can request location information from the Location Server located at the URI <http://lis.example.com/location>. The "Content-Type" and "Accept" headers indicate that this request contains a HELD XML structure, and the response to it should too. Since this basic request only contains an XML "locationRequest" element, the server can answer this request with either location by value or by reference, or with both forms of location. In addition, the server can provide a location value in either civic or geodetic form, that is, either as a pair of coordinates or as an address (or both). (Examples of these responses are shown a little further down.) In this case, the server is free to return any of these types of information in its request. Another example shows how a request can specify which forms of location the client would prefer:

```
POST /location HTTP/1.1
Host: lis.example.com
Accept: application/held+xml,
  application/xml;q=0.8,
  text/xml;q=0.7
Accept-Charset: UTF-8,*
Content-Type: application/held+xml

<?xml version="1.0"?>
<locationRequest
          xmlns="urn:ietf:params:xml:ns:geopriv:held">
  <locationType exact="true">
     geodetic
     locationURI
  </locationType>
</locationRequest>
```

In addition to the basic request in the previous example, the client can specify which types of location information (XML element "locationType") it would like to receive – in this case, coordinates ("geodetic") and a reference ("locationURI"). The "exact" attribute specifies that the server must return only the location types in the "locationType" element, and no others. For example, if the server had only an address for the client ("civic"), then it would have to return an error in response to the above example request. If the "exact" attribute had not been set (since the default value for this attribute is "false"), the server would not have to follow the client's preferences, and could return another type of information that it had available.

The Location Server's response to the above example request could take the following form:

```
HTTP/1.1 200 OK
Server: Example LIS
Date: Tue, 10 Oct 2010 03:42:29 GMT
Expires: Tue, 10 Oct 2010 03:42:29 GMT
Cache-control: private
Content-Type: application/held+xml

<locationResponse
            xmlns="urn:ietf:params:xml:ns:geopriv:held">
  <locationUriSet expires="2010-10-10T21:29:00+01:00">
    <locationURI>
      https://lis.example.com:1234/0815ax92383jf94
    </locationURI>
    <locationURI>
      sip:0815abc@lis.example.com
    </locationURI>
  </locationUriSet>
  <presence xmlns="urn:ietf:params:xml:ns:pidf:geopriv10"
           entity="pres:0815@10.10.0.9">
    <tuple id="loc">
      <status>
        <geopriv>
          <location-info>
            <gs:Circle
                xmlns:gs="http://www.opengis.net/pidflo/1.0"
                xmlns:gml="http://www.opengis.net/gml"
                srsName="urn:ogc:def:crs:EPSG::4326">
              <gml:pos>48.4 15.1</gml:pos>
              <gs:radius uom="urn:ogc:def:uom:EPSG::9001">
               100
              </gs:radius>
```

```
          </gs:Circle>
        </location-info>
        <usage-rules>
          <retransmission-allowed>
            true
          </retransmission-allowed>
          <retention-expiry>2011-01-01T10:10:44+01:00
          </retention-expiry>
        </usage-rules>
        <method>Wiremap</method>
      </geopriv>
    </status>
    <timestamp>2010-10-10T04:42:29+01:00</timestamp>
  </tuple>
 </presence>
</locationResponse>
```


As the client requested, this response provides location information in the form of coordinates (with an error radius, in the "Circle" element) and a location URI (in the "locationURISet" element). The "presence" element contains the full PIDF-LO document, which could be used, for example, in a SIP request. The "expires" attribute of the "locationURISet" is important: It specifies the time after which the URIs in the set are no longer valid, in the sense that they can be used to request location information (i.e., it specifies how long the Location Server will respond to queries for these URIs).

If a client requests several types of location information with the "exact" attribute set to "true", the server must return all of them. So if the client requests "civic" and "geodetic" location types, but the server has only one of the two types, then it must return no location information, answering instead with an error message. This structure means that there is not a simple way for the client who wants to receive location by value (but doesn't care whether it receives coordinates or an address) but not a location URI: If the client requests both "civic" and "geodetic", with "exact" set to "true", then the server must respond with an error if it has only one type (even though the client would have accepted either). Clients can work around this problem by sending two separate queries for civic and geodetic, or by setting the "exact" attribute to "false".

In addition to the "civic" and "geodetic" location types, the client can use the "locationURI" location type to request location by reference, or the "any" type to specify that the server can return any type of location.

An error message in a HELD response has the following form:

```
HTTP/1.x 200 OK
Server: Example LIS
Date: Tue, 29 Jan 2011 11:20:00 GMT
Expires: Tue, 30 Jan 2011 11:20:00 GMT
Cache-control: private
Content-Type: application/held+xml

<?xml version="1.0"?>
<error xmlns="urn:ietf:params:xml:ns:geopriv:held"
    code="locationUnknown"
    message="Unable to determine location"/>
```

This error message indicates that the Location Server is not in a position to determine the location of the client.

Full descriptions of all protocol parameters, as well as some other HELD examples, can be found in the full HELD specification (RFC 5985, Barnes et al., 2010). The IETF GEOPRIV working group is currently working on a few HELD extensions.

Before a client can even send a request to a Location Server, there is the question of how the client finds a Location Server to contact. Location Servers usually cover a specific geographical area (since they have to be able to determine a client's location), so it's essential to contact the correct Location Server. A Location Server in a different network far away from a client typically cannot determine the client's position.

HELD thus has a mechanism for networks to inform clients about which Location Server they should contact (RFC 5986, Thomson and Winterbottom, 2010), or, in other words, a way to clients to discover the Location Server recommended by their local network. The discovery process has two steps: First, the client receives a DHCP option that contains a domain name for the local network. Second, the client sends a DNS query to find all Naming Authority Pointer (NAPTR) records for that domain. The URI for the Location Server is contained in a NAPTR record of the following form:

```
example.com IN NAPTR 100 10 "u" "LIS:HELD"  \
            "!.*!http://lis.example.com/location!".
```

This example record says that the HELD Location Server (indicated by the service tag "LIS:HELD") for the domain "`example.com`" can be accessed by sending queries to the HTTP URI `http://lis.example.com/location`. The network can advertise multiple Location Servers by including multiple NAPTR records under the same domain name. If there are multiple records, then the client uses the "order" and "preference" values in each record (in this case, set to "100" and "10", respectively) to determine the order in which to try the servers, as defined in the U-NAPTR specification (Daigle, 2007).

The Location Server advertised through this process is the one that the network believes will provide good information; the client is of course free to use another server.

4.1.2 DHCP Options for Location Configuration

The Dynamic Host Configuration Protocol (DHCP) is used in many networks throughout the Internet today for configuring client hosts (e.g., assigning IP addresses and DNS resolvers). So it makes sense to configure hosts with location information via DHCP as well. Two DHCP extensions ("options" in the DHCP terminology) have been defined to carry location information:

1. Option 99, for location information as an address (RFC 4776 (Schulzrinne, 2006)).
2. Option 123, for location information as a set of coordinates (RFC 3825, Polk et al., 2004), to be updated by (Polk et al., 2010).

Both DHCP options can be configured into a DHCP server. At the time of writing, DHCP only supports location by value (using the options described here), but work is in progress to add support for location by reference (Polk, 2010).

The format of DHCP option 123 is shown in Figure 4.2, and the individual elements are briefly explained in Table 4.1. Likewise, DHCP Option 99 is described in Figure 4.3 and Table 4.2. Here we do not make a distinction between DHCP (used with IPv4) and DHCPv6 (used with IPv6), because the format of the options is the same for each (technical details can be found in RFC 3825 and RFC 4776).

If a DHCP server is responsible for only a single, geographically-constrained part of the network (e.g., a single building), it can be enough to simply statically configure values for these DHCP options on the server. All clients will thus be configured with the same location information (e.g., the same address for the building). In this case, the "what"

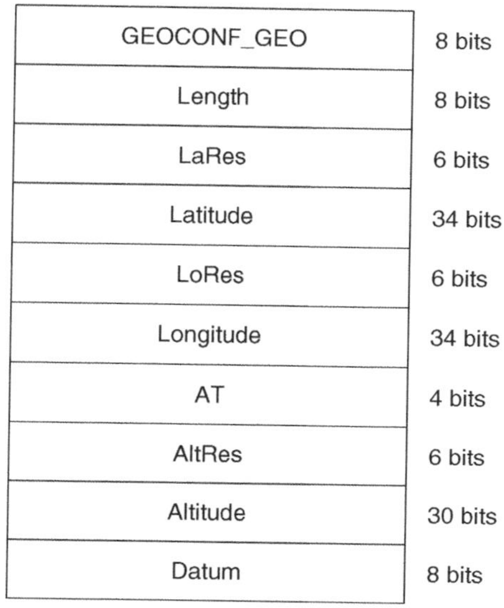

Figure 4.2 DHCPv4 Option 123 for location configuration via coordinates.

Table 4.1 Elements of DHCPv4 Option 123

Field	Contents
Code	123 (the fixed code for this DHCPv4 option)
Length	16 bytes (the fixed length of this DHCP option)
LaRes	Resolution of the latitude value (integer, in bits)
Latitude	Latitude value (fixed point, 9 bits integer / 25 bits fraction)
LoRes	Resolution of the latitude value (integer, in bits)
Longitude	Longitude value (fixed point, 9 bits integer / 25 bits fraction)
AT	Altitude type (e.g., meters vs. floors)
AltRes	Altitude resolution (integer, in bits)
Altitude	Altitude value (format defined by AT field)
Datum	Coordinate reference system (e.g., value 1 indicates WGS84)

element of the value for option 99 should be set to “0” to indicate that the location of the DHCP server is being provided; no corresponding value is defined for option 123.

Some other common problems encountered when configuring the civic address elements in option 123 are discussed in Section 3.2.1.

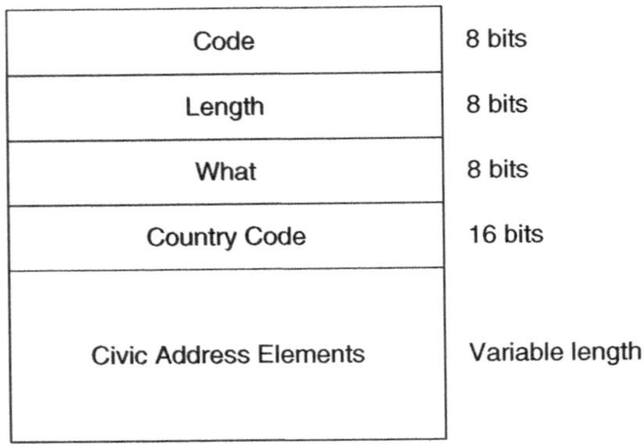

Figure 4.3 DHCPv4 Option 99 for location configuration via civic address.

Table 4.2 Elements of DHCPv4 Option 99

Field	Contents
Code	99 (the fixed code for this DHCPv4 option)
Length	The length of this DHCP option
What	What is located at this address (the endpoint, the nearest network element to the endpoint, or the DHCP server)
Country Code	Two-letter country code for this address (from ISO 3166)
Civic Address Elements	Address elements (e.g., city, street name, house number)

If a network uses DHCP for client configuration anyway, a client can obtain an IP address (and all the things it would normally get) delivered alongside location information.

4.1.3 LLDP-MED

The Link Layer Discovery Protocol (LLDP) was standardized by the IEEE (IEEE Std 802.1AB (*Station and Media Access Control Connectivity Discovery*, 2009)), as a layer-2 configuration mechanism. So-called "LLDP Agents" can share information between neighboring devices in the network. Although LLDP is an extension of the Cisco Discovery Protocol (CDP), it is not a vendor-specific protocol. The Media Endpoint Discovery extensions to LLDP were defined by TIA (Standard ANSI/TIA-1057, *Link Layer Discovery Protocol for*

Media Endpoint Devices, 2006); with these extensions, we refer to a collective "LLDP-MED" protocol. LLDP-MED can be used for location configuration in support of emergency calling as well as non-emergency location-based services. LLDP-MED offers three types of location information:

- coordinates (in the same format as DHCP option 123);
- addresses (in the same format as DHCP option 99);
- Emergency Location Identification Number (ELIN).

For location in the form of coordinates or addresses, the LLDP-MED formats are the same as the DHCP formats discussed in the previous section. In addition, LLDP-MED has a third possibility, ELIN. An ELIN value is a valid E.164 telephone number, which many major PBX products use to identify the source of an emergency call. ELIN is widely used in current PBX products in North America, but can also be used for location configuration in general, although in order to be useful, ELIN values distributed in LLDP-MED must clearly be bound to some determined location. (e.g., currently in Austria, all geographical telephone numbers are tied to a fixed network access point.)

However, it is important to note that the use of ELIN for location configuration is not generally supported in the ECRIT emergency calling architecture. This is because the concept of ELIN comes from the old telephone system (the PSTN) and the IETF is focused on a new emergency calling architecture for the Internet, without much consideration being given to transition scenarios or improvements to the current PSTN model. This situation leads to a problem, in that a client could obtain valid location information in the form of ELIN that cannot be used to make an emergency call, since it does not make sense in the Internet context. For this reason, configuring hosts with ELIN is not advised.

Compared to the protocols described above, LLDP-MED has one other peculiarity: Location information is delivered to clients in periodic, unsolicited announcements, in contrast to HELD and DHCP, where the client must first send a request to the server. This feature has the advantage that the client does not need to know or discover the address of the Location Server. A small disadvantage of LLDP-MED is the limitation that only one type of location information can be configured, for example, coordinates or an address, but not both. However, the IEEE is considering some extensions to LLDP-MED in order to allow the configuration of different types of location information at the same time. In addition, LLDP-MED does not currently support location by reference. If such support were desired, it could be done in a similar fashion to DHCP, where a solution is currently being developed (Polk, 2010).

LLDP frames are always addressed to the same destination MAC address, the LLDP Multicast Address (`01-80-C2-00-00-0E`). Currently-available switches do not handle these packets in a uniform manner; some drop them and some forward them to all ports.

However, note that LLDP-MED is not mandatory in the ECRIT architecture. Hence, every network has to support at least HELD or DHCP for location configuration and may optionally choose to offer LLDP-MED if suitable.

4.1.4 Protocol Comparison

Since we've described the different possibilities for location configuration – HELD, DHCP, and LLDP-MED – their important features and differences are summarized in Table 4.3. According to the current documents describing the ECRIT architecture, network operators must support at least one of the two IETF protocols HELD and DHCP, LLDP-MED support is optional. End devices must support both HELD and DHCP, so that they can perform location configuration in whichever network they use. Additionally, end devices may choose to support LLDP-MED.

4.1.5 Conversion between Location Formats

In order to be sent in an emergency call, location information must be formatted as a PIDF-LO document. Of the three location configuration protocols, only HELD provides the client with a PIDF-LO document; location information in the DHCP format (which is also used in LLDP-MED) must be converted into the PIDF-LO format before it can be used in an emergency call.

Addresses are specified with civic address elements in both formats, as binary type/length/value elements in DHCP/LLDP-MED and as XML elements in PIDF-LO; the translation between the two is specified in RFC 4119 and RFC 5139 (Peterson, 2005; Thomson and Winterbottom, 2008). With coordinates, the translation is somewhat more difficult, because the DHCP format includes uncertainty information. If one is willing to throw away this uncertainty information, the coordinates from the DHCP format can be translated directly and

Table 4.3 Comparison of the HELD, DHCP, and LLDP-MED protocols for location configuration

	HELD	DHCP	LLDP-MED
Location by Value	Yes	Yes	Yes
Location by Reference	Yes	No	No
Server discovery required	Yes	No	No
Configures address and coordinates at the same time	Yes	Yes	No
Location update	Client must send another request, or subscribe to a location reference	Client must send another request	Periodic announcement of current location
ELIN	No	No	Yes
Possible deployment problems	NAT and VPN concerns		Inconsistencies in LLDP Multicast Address forwarding
Conversion required for SIP usage	No	Yes	Yes
Flexibility of location information	Full PIDF-LO profile	Highly constrained format	Highly constrained format
Appropriate for WLAN?	Yes, especially with measurements	Yes, with network-based positioning	Only provides same location to all endpoints

encoded as a point (i.e., a set of coordinates). If the uncertainty fields are to be preserved, then the resulting shape is a polygon; further details can be found in the update to RFC 3825 that is currently being developed (Polk et al., 2010).

Of course, a PIDF-LO document can accommodate both geographical location information and address data. As already mentioned, ELIN information from LLDP-MED cannot be translated into the PIDF-LO format.

Indeed, in addition to the location information itself, there are also other elements of the PIDF-LO that will need to be filled in. A summary of the important elements in a PIDF-LO was given in Section 3.2.1. For the most part, there is no standard or current practice for how these fields are populated; in the future, recommendations from the IETF or groups specifying deployment architectures could be helpful in ensuring consistent usage.

If a client supports multiple different protocols for location configuration, there is inevitably a question as to what order they should be used in. A recommendation from the Broadband Forum suggests the following order:

1. LLDP-MED.
2. DHCP.
3. HELD.

As long as one method is successful, the client can stop the location configuration process and not use any other protocols. Requests to refresh location simply make use of the last protocol used successfully. If this protocol fails for some reason, then the client tries the others.

4.2 Positioning Using GPS

By using a GPS receiver, a device can determine its own location. This clearly has the advantage that no protocol or server is necessary in order to make location information available to the device – the device is independent. Since GPS receivers have continued to get smaller and less expensive, they can already be found in many mobile telephones – why not in VoIP telephones too?

Of course, one must also consider the limitations of GPS, such as the difficulty in getting a signal in dense urban areas and the time that it takes to establish the receiver's position (the so-called "time to first fix"). Some of these problems can be mitigated with Assisted GPS

(A-GPS), but that returns to the need for an assistance server and a protocol to use with it.

If a device uses a GPS receiver to determine its own location, then it must construct a PIDF-LO document itself. As noted in the PIDF-LO profile, a circle can be used to represent coordinates with an uncertainty radius – as location is usually presented in GPS devices. The "dilution of precision" (DOP) value provided by the GPS receiver corresponds to the radius of the circle, but it must be converted into meters before being added to the PIDF-LO. If altitude information is also used, the ellipsoid profile can be used to also incorporate vertical uncertainty.

Location information determined via GPS can also be used for LoST mapping and included in SIP messages (using the location conveyance mechanism). LoST clients are supposed to compare their location to the "serviceBoundary" element of the LoST mappings they have received, in order to determine whether they are still within the jurisdiction of the PSAP in the mapping. With this information, the client only needs to update its mapping information if it leaves the service boundary indicated by its current mapping. (This process is described in Section 3.4.4.) In the context of location conveyance, there may be a need to provide updated location information. In this case, the device must have the capability to provide current location information to the PSAP, either by sending an update or by implementing a simple Location Server. This latter option would offer the advantage that the PSAP could subscribe to the client's location and specify the conditions under which it would like to be notified (e.g., requesting an update if the location changes by a certain number of meters). The need for this functionality clearly increases the complexity of the end device.

In any case, location determination via GPS represents a meaningful complement to location configuration.

4.3 Network-Based Positioning

In order to be able to provide a location configuration service to clients, a Location Server has to have a way of determining where clients are located. For this purpose, each type of network requires a different set of techniques – so there are more possibilities than could be described in this chapter. In any case, it is important that each network operator consider how it is possible to locate subscribers in his network.

For fixed networks, "positioning" using the layout of the cables in the network can be a useful technique. Wire maps frequently provide high-precision location information, such as the building, floor, and room where a wire terminates. Based on knowledge of what switch and port a subscriber is connected to, one can look up location information from a wire map. A Location Server can obtain this information by requesting managed switches using the Simple Network Management Protocol (SNMP).

For Internet Service Providers operating over a fixed network, databases of subscriber addresses (e.g., service addresses) can be a good foundation for providing location information. To use subscriber addresses as a source of location information, the Location Server must be able to identify which subscriber has sent a given query for location (e.g., mapping the source IP address of a request to the subscriber to whom it has been assigned), so that it can retrieve that customer's address information from the customer database.

In wireless networks, the address of the base station to which an device is connected can be helpful (as long as the area served by the base station is not too large). Obviously, it is better to locate the device itself, for instance, using triangulation, where this is feasible. Some brief discussion of techniques used in different types of wireless networks is included in the ECRIT framework document (Rosen et al., 2009).

Of course, several network equipment vendors already offer products for location subscribers within a network, and many networks already use these products to be able to locate their subscribers, for example, for E9-1-1. The location information provided by these products can of course be used as a source for a Location Server.

With each of the different possible positioning mechanisms, one must take into account the time required to establish a position, in addition to the possible impacts on the network (e.g., additional traffic on the network generated by SNMP queries).

4.4 Location Hiding

"Location hiding" is currently used in the IETF to describe scenarios in which a location provider (e.g., an ISP operating a Location Server) wishes to prevent endpoints from knowing their precise location. As a reminder: In the ECRIT architecture, the very first step in placing an emergency call is for the endpoint (i.e., the VoIP telephone) to determine its location, so that it can perform the other steps in the calling process, namely LoST mapping and location conveyance. It is possible for a VoIP operator to determine and introduce location information as part of the emergency call setup, but even in this case, the telephone itself still needs some location information to even recognize that an emergency call has been placed (based on a local emergency dial string, see Section 3.5.1).

Some telecom operators, however, have expressed the opinion that providing location information to devices connected to their networks raises both commercial and privacy concerns. For example, location information that the network provides for emergency calling purposes can also be used for other location-based services. So while the network operator bears the cost of deploying and operating the infrastructure to provide location, other service providers can benefit from this infrastructure for free. How can the network operator arrange things so that location information is available for emergency calls for free but for other services it is only available for a fee? Solving this problem requires preventing the user from accessing location in some cases, that is, hiding the location. (Further information on location hiding problems can be found in the ECRIT location hiding requirements document (Schulzrinne et al., 2010)).

One possible solution would be for the location provider to only provide location by reference. PSAPs could send requests for the reference for free, while other service providers could pay for access. However, in order to determine the responsible PSAP (along with dial strings, etc.), the LoST infrastructure requires location by value – a location reference cannot be used. On the other hand, the LoST mapping process often does not require *precise* location information. Frequently, information about the country or province where the user is located is enough to enable the LoST server to determine the proper PSAP. So one solution to the "location hiding problem" would be for the network to always use the HELD protocol to configure endpoints with *both* location by value (representing the approximate location of the endpoint, to determine the proper PSAP) and location by reference (which the PSAP can use to access precise location information). A telephone that has both a location value and a location reference must send both of them to the PSAP, so that the PSAP has access to both the rough and precise location information. This solution is currently under development in the ECRIT working group (Barnes and Lepinski, 2010).

However, it should be noted that the discussions of location hiding have implications for the how emergency calling architectures work, and location hiding solutions add additional complexity to emergency calls. It should never be the case that a PSAP is unable to obtain information about the location of a caller because it is not authorized by some access control policy put in place to hide the location. On the other hand, there is a risk that without location hiding, telecom companies will not make location information available at all.

4.5 Default Location

For cases where location determination isn't possible, the ECRIT architecture has a notion of a so-called "default location". In such cases, the Location Server – when it cannot determine the client's location – might return a pre-configured default location value to the client, for instance, the location of the client's ISP. In order to allow the client to recognize when this has happened, these default location values must be specially marked. This marking can be done in the PIDF-LO format with the "method" element, which should be set with the value "Default" – although this method token has not yet been registered with IANA.

The end device can use this default location to perform LoST mapping. Assuming that the same emergency services are available at the default location and at the caller's actual location, this will at least ensure that the device will be able to recognize when emergency calls are placed. In addition, an endpoint using a default location will only be able to contact the PSAP that is responsible for the default location. However, since the location information is marked as default location, the PSAP can recognize that the caller may actually be located somewhere else. That is, the default location can at least help the caller to reach *some* PSAP, even if it isn't the correct PSAP – which is better than not being able to reach any PSAP at all.

References

Barnes M, Winterbottom J, Thomson M and Stark B (2010) HTTP Enabled Location Delivery (HELD). RFC 5985.

Barnes R and Lepinski M (2010) Using Imprecise Location for Emergency Context Resolution. Internet Draft (work in progress) draft-ietf-ecrit-rough-loc.

Daigle L (2007) Domain-Based Application Service Location Using URIs and the Dynamic Delegation Discovery Service (DDDS). RFC 4848.

Link Layer Discovery Protocol for Media Endpoint Devices (2006). ANSI/TIA.

Peterson J (2005) A Presence-based GEOPRIV Location Object Format. RFC 4119.

Polk J (2010) Dynamic Host Configuration Protocol (DHCP) IPv4 and IPv6 Option for a Location Uniform Resource Identifier (URI). Internet Draft (work in progress) draft-ietf-geopriv-dhcp-lbyr-uri-option.

Polk J, Schnizlein J and Linsner M (2004) Dynamic Host Configuration Protocol Option for Coordinate-based Location Configuration Information. RFC 3825.

Polk J, Schnizlein J, Linsner M, Thomson M and Aboba B (2010) Dynamic Host Configuration Protocol Options for Coordinate-based Location Configuration Information. Internet Draft (work in progress) draft-ietf-geopriv-rfc3825bis.

Rosen B, Schulzrinne H, Polk J and Newton A (2009) Framework for Emergency Calling Using Internet Multimedia. Internet Draft (work in progress) draft-ietf-ecrit-framework.

Schulzrinne H (2006) Dynamic Host Configuration Protocol (DHCPv4 and DHCPv6) Option for Civic Addresses Configuration Information. RFC 4776.

Schulzrinne H, Liess L, Tschofenig H, Stark B and Kuett A (2010) Location Hiding: Problem Statement and Requirements. Internet Draft (work in progress) draft-ietf-ecrit-location-hiding-req.

Station and Media Access Control Connectivity Discovery (2009). IEEE.

Thomson M and Winterbottom J (2008) Revised Civic Location Format for Presence Information Data Format Location Object (PIDF-LO). RFC 5139.

Thomson M and Winterbottom J (2010) Discovering the Local Location Information Server (LIS). RFC 5986.

5

Implementation and Regulatory Considerations

Emergency calling systems necessarily create a link between the commercial and government sectors. At base, this link arises from the basic goal of an emergency calling system: to enable subscribers to *commercial* telecommunications networks to reach *government* services that are needed in the case of an emergency. In addition, because emergency services are so critical to the safety and security of the communities they serve, they are frequently subject to systems of regulation that define which services need to be available to the community, and how those services are requested and delivered. Depending on the jurisdiction, these regulations can be created and administered by either government agencies or industry self-regulatory bodies, and they can define everything from local emergency dial strings to complex network and tariffing structures that ensure the reliability and economic sustainability of emergency services in a particular market.

Of course, regulations can vary from country to country, between states or provinces within a country, or even between cities. So one book cannot provide a very comprehensive overview of the global regulatory situation. Instead, this chapter looks at regulatory and implementation issues at two levels: First, we look at the general tasks that are necessary for VoIP emergency calling to be implemented in a jurisdiction, and who the logical organizations are to undertake these tasks. These practical implementation considerations, and the ambiguities that remain, highlight areas that regulations will need to take into account.

Then, we'll present a few case studies of how emergency calling is handled in a few jurisdictions around the world, and how some of these jurisdictions are adapting to support VoIP emergency calling. First, we take an in-depth look at the regulatory situation in Austria, as

VoIP Emergency Calling: Foundations and Practice Karl Heinz Wolf and Richard Barnes

an example of a nation with a mature telecommunications industry that still has a unitary regulatory regime. Next, we consider the United States and the European Union, as examples of large, heterogeneous regulatory systems. Finally, we look to Japan for an intermediate case, and a non-Western perspective.

We note that it may seem strange to treat Austria and the European Union independently, as we do here, since Austria is a member state of the EU. The Austrian emergency calling system, however, pre-dates the formation of the European Union, and the EU continues to delegate almost all of the implementation details of emergency calling systems to the member states. So the Austrian system is still an essentially autonomous, complete system, providing a detailed example of how a national emergency calling system can be implemented, and an example of the sort of existing systems that might play into an overall European framework.

5.1 Distribution of Implementation Tasks

In order for callers to be able to place emergency calls seamlessly over VoIP in the future, there are several challenges that multiple different organizations need to overcome. The standards defined for emergency calling (e.g., in the IETF) only specify the interfaces between the components of the system – not who has to operate which components. This section lays out how the challenges and responsibilities involved in the VoIP emergency calling system could be distributed. A similar summary of these responsibilities can be found in the IETF "phone BCP" (Rosen and Polk, 2010).

5.1.1 Emergency Call Centers (PSAPs)

Most PSAPs today are only reachable over the current telephone network, that is, the PSTN. The ECRIT emergency calling architecture lays out how PSAPs can be reachable via Internet telephony (since calls will need to be handled independently of the PSTN, especially VoIP calls). For that reason, it's important for PSAP operators to prepare themselves for emergency calls over VoIP. The improvements that will be necessary for VoIP will also make it easy for other types of media to be used. Video and instant messaging can complement traditional voice calls, enabling more effective emergency calls and helping call takers get a better understanding of the emergency situation. By integrating Internet technologies, PSAPs can also be well-positioned for changes in citizens' communication behaviors (e.g., the trend toward VoIP and instant messaging).

As a transition strategy, it's possible for PSAPs to operate gateways that enable VoIP calls to be answered at PSAPs that are only connected to the PSTN. This strategy has the benefit that it's not necessary for each PSAP to have its own gateway; rather, several PSAPs can share such a gateway. However, current VoIP-to-PSTN gateways aren't quite sufficient: Because VoIP emergency calls will arrive with location information attached, the PSAP gateway will have to cache that location information and make it available to the PSAP outside of the call signaling (e.g., by installing it in an ALI database), since the PSTN signaling that arrives at the PSAP cannot carry location information. In addition, new mechanisms will be required to support calls back from the PSAP to the caller, since not all VoIP users have telephone numbers. For example, in Japan, the gateway between the VoIP system and the PSTN manages this state, tracking users to support callback; this solution, however, works only if the user is not very mobile, for example, if the IP address of the connected endpoint does not change.

In any case, implementing the ECRIT emergency calling architecture can create a few improvements for PSAPs. The most important is that location information can be delivered directly to the PSAP, so that response planning can be made much easier.

5.1.2 VoIP Software and Hardware Manufacturers

Vendors of software or hardware for VoIP should be prepared to incorporate elements of the emergency calling architecture into future products, and in updates to their current systems. It should be noted that it's not only phones (or soft phones) that need updating, but possibly also some other elements as well. For example, in order for SIP proxies to handle emergency calls, they will need to handle multi-part SIP messages containing the usual SDP body as well as a PIDF-LO location object.

The IETF document on Best Current Practices for emergency calling (Rosen and Polk, 2010) documents detailed requirements for client and proxies. One point bears emphasizing again here: Clients have to support both of the IETF protocols for location configuration (DHCP and HELD, others optionally). A built-in GPS receiver can of course also be used, and it can still be useful to have an option for manual user configuration as a fall-back.

Throughout many parts of the emergency services community, having a big, red Emergency button on a phone is not recommended – such an interface would introduce too great a risk that an emergency call would be placed accidentally. Even in the future of VoIP calling, emergency calls should be placed by a caller deliberately dialing an emergency number. The main difference for VoIP is that rather the phone will need to discover local numbers through LoST (possibly in addition to being configured with some home emergency number).

For example, nic.at, through its project supported by the ZIT Center for Innovation and Technology, has extended the Zap! VoIP client to support location configuration, recognizing emergency calls (by discovering the local dial strings), and location delivery to the PSAP. Further information can be found in Chapter 6.

5.1.3 Network Operators and ISPs

Network operators have arguably the most critical task of all the players in the emergency calling system, since they have to ensure that their subscribers have access to location information. Information on the location of the caller is critical for an emergency call. Missing or incorrect location information can result in mis-routed or failed emergency calls, with potentially life-threatening consequences. On the other hand, many network operators already have location resources that can be reused for IP location services. For example, many IP access networks today are built using legacy PSTN infrastructure, including fixed-line and cellular networks. In most jurisdictions, these PSTN networks have needed a location function for PSTN emergency calls for some time; all they need to do to support IP emergency services is provide a way for this location function to be accessed by subscriber devices over standard protocols.

The responsibility for providing a Location Server falls to the network, since it has the most direct, physical information about subscribers. Network operators have to provide a location service in their network, but they have a choice of which protocol to use for location configuration (either DHCP or HELD must be supported). Of course, multiple protocols could be used (e.g., all three of LLDP-MED, DHCP, and HELD). The decision about which protocol to use will naturally depend on which protocols are already deployed in the network and how the network is structured. Likewise, different network operators will need to choose different ways of providing location information to the Location Server, for instance, from a customer database or by using a triangulation service. Network equipment vendors already offer partial solutions to the problem of locating subscribers in the network. Investing in creating a location service can actually pay for itself as an entry into the location-based services market. The discussion of location hiding (see Section 4.4) shows that location information is valuable.

In the current emergency calling system, information about callers' location is provided to PSAPs free of charge. If VoIP emergency calls are going to be provided to the caller free of charge, location information will need to be provided as well. For other services, however, the network operator may still charge for access to location information.

5.1.4 VoIP Operators

VoIP networks are actually not strictly necessary for emergency calls in the ECRIT architecture. If the caller can establish an IP connection to the PSAP, then he can send SIP signaling directly. However, it is necessary for VoIP operators to be prepared to play a part in the emergency calling system. After all, emergency calls can also run over servers operated by VoIP providers. VoIP operators may thus need to upgrade some of their servers, for example, to handle calls addressed to Service URNs, to not crash when they receive a multi-part SIP message, and to forward location information appropriately. As an additional service, VoIP operators could provide certain PSAPs with verified identity information about their subscribers, such as a home address, full name, or other contact information. A caller's name can be useful, for instance, if the subscriber has configured his VoIP phone with only his first name or an abbreviation. In addition, the VoIP operator will need to support call-backs from the PSAP, for example, disabling the forwarding of calls to another number or voicemail.

Another important task for VoIP operators is to inform their customers about emergency calling features. Currently, customers are often given no information at all about emergency calls, and sometimes they get misleading information. It would definitely be helpful for callers to have instructions on testing that emergency calls work (the ECRIT architecture describes a test function) or to know which SIP phones support emergency calling. In addition, VoIP operators might help by providing configuration information to subscribers, or even running a LoST server that customers' phones can use (which then forwards queries to the authoritative server).

5.1.5 PSTN Operators

PSAPs are currently connected to the public telephone network, that is, the PSTN. In the transition phase, when VoIP calls pass through a gateway onto the PSTN in order to reach a PSAP, PSTN operators will naturally have some involvement in VoIP emergency calling, particularly on the issues of routing emergency calls, charging for them, and giving them priority treatment.

There is also some potential for improving the handling of emergency calls in the PSTN. For emergency calls in many jurisdictions today, PSAPs cannot receive automatic location information, instead having to make a special request for it. Some of the same location interfaces that support VoIP emergency calling could also provide automatic location for PSTN emergency calls.

5.1.6 Unassigned Responsibilities

The emergency calling protocols in the IETF were designed so that VoIP emergency calls could work in basically the same way all over the world. However, there are still points that need to be elaborated at the national level, since the handling of emergency calls is mostly a local function. At the national level, there are thus a few tasks to assign:

- collecting data for the mapping information for a LoST server;
- operation of the LoST infrastructure;
- defining the mapping from the local address format to PIDF-LO (see Section 3.2.1);
- defining the transition strategy (e.g., the deployment of gateways to make current PSAPs available over VoIP);
- assigning Service URNs to current emergency services.

There is currently no clear answer on who will take on these tasks. The IETF and other standards bodies are of course unable to take into consideration the specific circumstances of each nation. Nonetheless, these problems must be addressed in order for the emergency calling system to function smoothly. Since all the different parts of the system need to work together (e.g., using the same address format), it can be

Table 5.1 Possible division of tasks (TBD means To Be Determined)

Task	Responsibility
Operating a Location Server	Network operator/ISP
Mapping server (LoST server) and forest guides	TBD
Mapping the local address format to PIDF-LO	All local stakeholders
Implementing client functions (location and LoST)	VoIP equipment vendors
Assignment of Service URNs to services	TBD
Maintaining routing information in LoST	TBD
Updating the VoIP infrastructure	VoIP operators
Making PSAPs available via VoIP	PSAP operators

valuable for regulatory bodies or industry consortia to coordinate the development of emergency calling systems. For example, in the US, NENA has traditionally played this role.

5.1.7 Summary

The technical possibilities for placing emergency calls over the Internet have changed. Regulatory authorities should monitor these new developments so that they can put in place a framework to implement a new and improved emergency calling architecture.

As an overview, Table 5.1 lists the most important tasks and responsibilities within the emergency calling architecture, as well as some additional duties necessary for the transition from the PSTN to VoIP (most importantly, the operation of gateways).

5.2 Austria

This section gives an overview of the regulatory situation for emergency calls in Austria. The relevant statutes for emergency calls are specified in the Telecommunications Act (Telekommunikationgesetz, TKG) and in the Communications Parameter, Fees, and Value Added Services Act (Kommunikationsparameter-, Entgelt- und Mehrwertdiensteverordnung, KEM-V). For VoIP operators, there are some guidelines issued by the Regulatory Authority for Broadcasting and Telecommunications (Rundfunk und Telekom Regulierungs-GmbH, RTR) that also address the topic of emergency calling.

In addition, this section will also briefly cover some recommendations for the routing of emergency numbers from the Austrian

Technical Coordination Working Group (Arbeitskreis für technische Koordination, AK-TK), a group in which Austrian network operators discuss current issues for communications networks and services. Also important for future development of emergency calling systems is the work at the European level, such as the early studies done by the Body of European Regulators for Electronic Communications (BEREC), discussed below.

5.2.1 The Telecommunications Act

The Telecommunications Act of 2003 came into effect on 20 August 2003, and has been updated twice since then. The current version is available on the RTR homepage at `<http://www.rtr.at/de/tk/TKG2003>`. Section 20 of the Act deals with emergency calls, and makes it clear that:

- Operators of public telephone networks or telephony services are responsible for providing access to all emergency phone numbers.
- This access must be free of charge.
- The caller's telephone number must be available to the PSAP as identification.

If an operator does not fulfill these obligations, it is an administrative offense punishable with a fine of up to 37,000 euros (see Section 109, clause 3).

Section 98 describes the information that has to be provided to emergency service operators: Customer information (name, address, etc.) as well as location information can be provided if an emergency is in progress that justifies it. The emergency service operator must document this request within 24 hours, but the information is transmitted to the operator without delay.

The Telecommunications Act also defines the single European emergency number 112, requiring that information about the existence of this number be included in the common terms of service provided to telecommunications subscribers (see Section 25, clause 4).

5.2.2 KEM-V

In this regulation, there are a few more paragraphs that affect emergency calls. This regulation was issued by the RTR in 2009, it is available on the RTR web site at `<http://www.rtr.at/kem-v>`.

Table 5.2 Dial strings for all Austrian emergency services, according to KEM-V §17

Number	Service
112	Universal European emergency number
122	Fire department
128	Emergency number for natural gas pipe breaks
133	Police
140	Mountain rescue
141	Physicians on call
142	Counseling
144	Ambulance
147	Counseling service for children and youth

In the KEM-V, the Telecommunications Act's recommendations regarding emergency calls are made more concrete (e.g., intended applications, allocation of responsibilities, and rules of conduct). The regulation calls for the caller's phone number to be forwarded to the PSAP. If no phone number is assigned to a mobile telecommunications device, no number must be sent.

In addition, the KEM-V defines the short dial strings for emergency services and their meanings, which are listed in Table 5.2.

Standards of conduct for assigned emergency service operators are laid out in Section 19. This includes, for example, an obligation to ensure the availability of the service around the clock, throughout the area of responsibility. There is also a responsibility to ensure that callers are given adequate assistance for their emergency situation (e.g., transferring a call to a recorded message is not allowed). In addition, operators must provide routing destinations in a digital format. The KEM-V 2009 now also requires that all operators of emergency services forward emergencies outside their jurisdiction or requiring a different service (e.g., a caller in need of the fire department calls the ambulance) either by directly forwarding the emergency call or by recording all the data from the caller and forwarding the information about the emergency situation to the correct PSAP.

5.2.3 RTR Guidelines for VoIP Operators

The RTR conducted some research on VoIP, and released some guidelines on the topic, *Richtlinien für Anbieter von VoIP Diensten* (2005). In these guidelines, the RTR summarizes its position for VoIP operators, on the basis of the Telecommunications Act and the KEM-V. They deal not only with emergency calls, but also issues such as numbering, competition, interconnection, data retention, and lawful intercept.

The guidelines describe an important classification of current VoIP operators. Two classes of operators have been defined by the RTR, namely "Class A" and "Class B" VoIP services. "Class A" VoIP services are those that allow connections into and out of the PSTN, typically provided through a PSTN gateway and/or by assigning a telephone number to a subscriber. It is irrelevant where the gateway is located, or whether it is only rented by the operator – the only criterion is whether the service is offered in Austria. The RTR calls "Class B" VoIP services "Internet only" VoIP operators. With these services, only conversations with other VoIP subscribers are possible, and not connections to the PSTN. This classification is fundamental to the current question of whether emergency calls have to be supported or not. "Class A" operators must enable emergency calls, while "Class B" operators are not subject to this obligation, since they do not provide a service equivalent to the traditional public telephone service.

The RTR has pointed out, however, that its regulations can be quickly changed as needed (as might be the case if there are negative experiences with emergency calls). They have also emphasized that operators should make an effort to provide their customers with the usual level of access to emergency services.

In addition to the guidelines, the RTR also publishes a Frequently Asked Questions (FAQ) document. For example, in this document, they describe how one can direct an emergency call to the proper PSAP based on the subscriber's location. In that process, the location information is determined based on the caller's area code (this mapping is defined in an appendix to the KEM-V). The area code is then placed in front of the emergency number and the call is sent over a connection to Telekom Austria, who forwards the call to the responsible PSAP (see also Section 5.2.4).

The RTR issued their guidelines in October 2005 and is expected to issue a revision some time in the near future. Such an update could be triggered by subscribers' difficulties, changes in the market, or new ways of handling emergency calls over VoIP.

5.2.4 AK-TK Recommendations

In AK-TK, recommendations have been defined (in AK-TK recommendation EP 011) for transferring specific numbers in the "1" range – thus, all numbers that start with "1". This class of numbers covers all emergency calls (see EP 011 Section 3.5). Routing has to be based on the origin of the call. Two ways have been defined for how emergency calls can be delivered: In the first variant, the caller's area code is prepended to the emergency number. This provides an indication of which local network the caller is connected to, and indicates that the destination network must determine the proper PSAP for the call. In the second variant, the area code of the responsible PSAP is prepended. The destination network then delivers the emergency call to the corresponding PSAP.

A further AK-TK recommendation deals with the "Concept of Priority for Emergency Calls" (EP 003). In normal telephone networks, an emergency call can be recognized by an emergency number. A transit network can likewise examine the destination number for a call to determine whether it is an emergency call. In Austria today, all emergency calls are transferred to the Telekom Austria network, since all PSAPs are connected to this network. From the emergency number and the caller's area code, Telekom Austria can determine the geographical calling number of the responsible PSAP. Should there be a lack of capacity in the network, there are circuits (i.e., telephone lines) in so-called "emergency call bundles" available at interconnection points to carry emergency calls. These circuits are reserved exclusively for emergency calls, and are only used when other circuits (in the "normal bundle") are already in use. This arrangement is especially useful during catastrophic events.

5.2.5 Emergency Calling in Austria

This section describes how emergency calls are currently handled in the traditional fixed and mobile telephone networks in Austria. In Austria all PSAPs are connected to the Telekom Austria network, so all calls have to be routed through that network.

5.2.5.1 Fixed Networks

Because the telephone lines in the network are fixed in this case, the "positioning" of the subscriber is easy to accomplish: If the caller's telephone number is sent to the PSAP, checking a directory suffices to find the corresponding address. Thus, many PSAPs in Austria simply rely on a CD with the phone book on it as their mechanism for "positioning" callers (Stastny and Merka, 2006). This solution is naturally not optimal; for instance, unlisted numbers aren't on the CD and updates to the CD are only delivered periodically. There is sometimes a delay of a few weeks before the PSAP is aware of new telephone numbers or changes of address. In some cases, the PSAP can call the operator to get the information it needs, but such calls result in a delay in the handling of an emergency call and introduce additional risks that an error will occur.

The responsible PSAP is determined using the caller's area code. However, the regions that PSAPs cover are not always identical to the regions that correspond to area codes. So even in fixed networks, there is room to improve emergency call routing.

Nonetheless, the fixed network provides high availability. As of 2007, Telekom Austria estimated their availability at 99.96%, with emergency services available 99.99% of the time (*Qualitätskriterien für den Universaldienst gemäß Universaldienstordnung*, 2007).

In addition, "old-fashioned" analog telephones have an unusual advantage: They work even during a power outage, since the power for the end device comes from the telephone line. This benefit should not be underestimated.

Since the calling number for an emergency call is always sent with the call (even if the subscriber does not want it to be sent), the PSAP can place a return call to the caller easily.

Large enterprises with several locations can sometimes run into unique problems with emergency calling. If the enterprise's telephone system signals the main number for headquarters as the origin of an emergency call placed from a branch, this can lead to a delay in help arriving, since the emergency call would be routed on the basis of an incorrect area code. Placing a return call can also be difficult if the PSAP cannot connect directly back to the caller, especially if the number that is provided to the PSAP belongs to some central office that knows nothing about the emergency. In addition, telephone systems need to take into account how emergency calls can be set up in the first place. In many telephone systems it is common for the caller to have to dial a zero to get an outside line. How are emergency calls handled in this case? Does the caller also have to dial zero for these? If so, how does the caller know to do this? What if a visitor wants to place an emergency call? Moreover, one needs to consider whether the telephone system recognizes emergency calls (based on short dial strings).

Finally, in some enterprise phone systems, the common emergency numbers (122, 133, 144, etc.) are "overlaid" with direct-dial extensions. If an employee tries to place an emergency call, instead of a PSAP he only reaches one of his colleagues. These problems become even more worse with so-called "Virtual Private Network" products offered by mobile operators, by which an employee of a corporation can also be reached on his mobile telephone by simply dialing the same extension as on the office phone. A colleague of an author of this book experienced this problem in reality: For a while, he was occasionally interrupted by "emergency calls" from his colleagues at the weekends, because his extension within the office phone system was "123" – used otherwise to reach the Austrian towing service ARBÖ.

In the worst case, such an overlay can mean that an employee is completely unable to place an emergency call, since all emergency numbers lead to colleagues' extensions. In dialing plans within an enterprise, none of the local emergency numbers – or even better, no numbers starting with "1" – should ever be used for extensions.

Delivering Emergency Calls to Telekom Austria

In Austria, all emergency calls have to be transferred to the Telekom Austria network, since all PSAPs are connected to that network. Telekom Austria's interconnection agreement describes "Regulations for public short dial strings for emergency services" (see Section 16 (*Zusammenschaltungsvertrag*, 2008)). This document refers to the AK-TK recommendations, which are described in Section 5.2.4 above. Information on where an emergency call should be routed can only be found based on the caller's area code. If no area code is available, the emergency call should be routed to the PSAP responsible for the transfer point.

If in the future, local numbers in Austria are made more flexible, that is, if the area code no longer provides information about the caller's location, then there will be a problem providing correct handling for emergency calls. Thus it can be seen that even emergency calls in fixed networks can benefit from improvements to the emergency calling system that are developed to support VoIP. This book, however, is focused only on the VoIP emergency calls themselves.

Interconnected networks have two options for providing Telekom Austria with "positioning" information on fixed network subscribers: A network must provide a subscriber directory or a contact number where there is always someone who will provide the necessary information to the PSAP. Telekom Austria accepts changes to this information only at the beginning of each month. The freshness of location information provided in this way thus cannot be guaranteed.

Emergency calls are of course free of charge to the caller, but there are still costs that arise for the network operator. Telekom Austria charges each network operator between 0.71 and 2.25 euro cents per minute for emergency calls (depending on the time of day and the connection point). (This information is as of 2008, see page 98 of the interconnection agreement *Zusammenschaltungsvertrag* (2008).) For the service of determining the appropriate PSAP based on the area code for the caller's location, the operator pays a lump sum of 760 euros per month (also as of 2008). Operators are also charged by Telekom Austria for updating contact information or providing a new subscriber directory.

5.2.5.2 Mobile Networks

Due to the widespread use of mobile phones today, many emergency calls are placed via mobile phones. In Austria, 7.4 million emergency calls were placed from mobile phones in the year 2009. (The Austrian Mobile Communications Forum publishes detailed statistics on emergency calls at `http://www.fmk.at/content.php?id = 146`. Some recent statistics are shown in Figure 5.1.) Because subscribers are mobile, locating callers and routing emergency calls correctly can be problematic. Above all, the caller's phone number cannot be used to infer the caller's location, as is commonly done in fixed networks.

The network typically determines the responsible PSAP based on the base station that the caller is using. However, a base station within one PSAP's service area can support subscribers that are served by another PSAP. Especially near borders, the fact that the PSAP is chosen based on the base station's location, not the user's location, can cause callers to be routed to the inappropriate PSAP. When a network receives an emergency call, it transfers it to the Telekom Austria network with the area code of the base station prepended. The PSAP will be presented with the caller's number. If the caller's location is needed, the PSAP will need to request it from the mobile operator (this is mostly accomplished by sending a fax to the mobile operator). The mobile operator can then provide location information, however, this

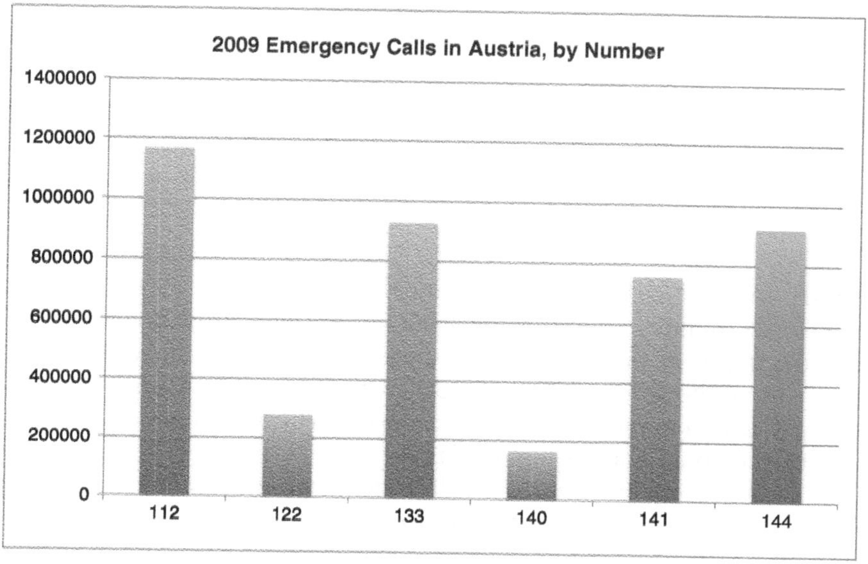

Figure 5.1 Mobile phone emergency calling statistics for 2009 in Austria. From the Mobile Communications Forum Emergency call statistics *7 Millionen Argumente fuer Mobilfunk-Infrastruktur* (2010). Current statistics at `http://www.fmk.at/content.php?id = 146`.

information can sometimes be very imprecise. Mobilkom Austria points out in its brochure *Handy Am Berg* (2006) that location information can differ from the caller's true location by up to a kilometer, and thus is not accurate enough to locate the caller precisely. Of course, positioning only works if the mobile phone is turned on. So one cannot necessarily rely on the positioning of a mobile phone in an emergency.

Furthermore the process of requesting location information by fax introduces some delay in emergency call handling. The facility to automatically send location information – or at least an automated way to request location information – would certainly result in time savings. In Switzerland, for example, a system is being deployed in which mobile operators automatically send location to a central system in the course of an emergency call, so that PSAPs no longer need to interact with mobile operators (*Leitweglenkung und die Standortidentifikation der Notrufe*, 2005).

Transferring a number from one operator to another can introduce further delays. Since only the caller's phone number is displayed to the

Table 5.3 Comparison of emergency calling in fixed and mobile networks

	Fixed	Mobile
Location	Subscriber directory	Fax to mobile operator
Routing	Caller's area code	Location of the base station
Call-back	Possible	Possible, if the device has a SIM

PSAP, in some cases the current mobile operator for a caller with a transferred number cannot be determined. In such cases, it can happen that the improper mobile operator is asked for location; since this operator will not be able to locate the caller, this operation wastes valuable time. A central number portability database, which has not been created in Austria, could correct this situation.

It is frequently possible for callers to place emergency calls over networks with which their home network has no roaming agreement. In Austria, emergency calls are even possible without a SIM card. It should be noted, however, that mobile phones (depending on the manufacturer) frequently have well-known emergency numbers pre-configured (e.g., 112 and 911), so that they can recognize emergency calls. However, other than the European emergency number 112, no other Austrian emergency numbers will work. In addition, phones without SIM cards cannot receive call-backs from the PSAP. Call-backs can fail in other cases as well, for instance, if the caller has forwarded incoming calls to another number or to voicemail.

So there are still some opportunities for mobile networks to make emergency calls more effective and to better support PSAPs in their daily work, especially by providing automated access to location information. In closing, Table 5.3 provides a brief overview of emergency calling in fixed and mobile networks in Austria.

5.3 The United States

In contrast to Austria, which has a single regulatory authority and a central telecom carrier, the situation in the United States is much more diverse. There are three levels of government in the US – federal, state, and local – and all three are involved in the regulation, funding, and management of emergency calling systems. The relationship between jurisdictions and telecom carriers is many-to-many, with each jurisdiction potentially being served by many carriers, and most carriers spanning more than one jurisdiction. Possibly the only thing that is consistent through the entire system is the single, unified emergency number, the well-known "9-1-1".

In this section, we will provide a brief overview of the history of 9-1-1 and the different organizations involved in regulating emergency calling, and in particular how location information is handled in current systems.

5.3.1 9-1-1 Regulation

Generally speaking, there are three "layers" of government in the US: federal, state, and local. At each of these layers, there are organizations that play a role in regulating emergency services. The day-to-day operations of PSAPs and the telecommunication networks they connect to are typically matters for local or municipal governments, with states sometimes setting state-wide standards for how emergency services are delivered. At the federal level, government organizations set high-level standards for emergency access. Also at the national level, but outside the government, the non-profit National Emergency Number Association (NENA) acts as a coordinating body among all these entities on both technical and policy matters.

At the highest level, US policy with respect to emergency calling is set by legislation passed by the US Congress. Some of the most important statutes are the following:

- **Communications Act of 1934** and **Telecommunications Act of 1996** – Establish the framework for telecommunications regulation in the US, most importantly through the creation of the Federal Communications Commission (FCC).
- **Wireless Communications and Public Safety Act of 1999** – Defines 9-1-1 as the single emergency number, and calls for parity between emergency services for wired and wireless callers.
- **ENHANCE 911 Act of 2004** – Establishes grants of up to 250 million dollars per year, administered by a National E9-1-1 Implementation and Coordination Office.
- **New and Emerging Technologies 911 Improvement Act of 2008 (the "NET 9-1-1 Act")** – Requires IP-enabled voice services to provide 9-1-1 services equivalent to wired and wireless telephone networks, and empowers the FCC to make regulations related to IP 9-1-1.

As these examples show, the overall role of the federal government in emergency calling is to set the overall requirements (e.g., whether 9-1-1 must be available for mobile or VoIP networks) and provide some funding to 9-1-1 operations. It should be noted, that the grants issued under these Acts are typically time-limited, and thus more for short-term projects such as improvements to PSAPs rather than for ongoing operation of PSAPs. PSAP operations and local coordination with telecom operators are essentially the domain of the states.

The Federal Communications Commission (FCC) is the federal regulator for all telecommunications in the United States. It was originally created by the Communications Act of 1934, as a successor to the Federal Radio Commission, and deals with all aspects of telecommunications, from radio spectrum licensing to ensuring equitable broadband access. As part of this role, it was the FCC that originally instituted 9-1-1 on a national level, and it continues to create regulation to update 9-1-1 requirements as the telecom marketplace evolves.

As the above-mentioned statutes show, as telecommunications have moved from wired networks to wireless networks, and now to IP-based systems, Congress has successively required that each of these networks provide the same level of access to emergency services. In step with these extensions of 9-1-1 – certainly after a law has been passed, but sometimes also before – the FCC defines rules for which communications systems are required to provide 9-1-1 services, and what sort of service must be provided. Typically the FCC does not, however, specify *how* 9-1-1 is to be implemented.

For example, in June 2005 (three years before the NET 9-1-1 Act), the FCC issued its now well-known ruling on VoIP emergency services (*IP-Enabled Services E911 Requirements for IP-Enabled Service Providers* 2005), which says that "interconnected Voice over IP" providers must provide E9-1-1 services "as a standard feature of the service". The initial notice made the above requirement firm, but sought feedback from the community on a reasonable implementation timeline, and allowed an interim solution where users report their own location.

The actual implementation of 9-1-1 services is left to telecommunications providers on the calling side, and to state and local governments, on the PSAP side. As suggested in our summary of relevant statutes above, the federal government does *support* the states in this effort, by supporting national coordination of implementation efforts and providing funding, but states typically make their own autonomous decisions about how to implement 9-1-1 services in their jurisdictions.

As a result, different states implement 9-1-1 in different ways, sometimes multiple ways within a single state. In the State of Indiana, for instance, wired and wireless carriers are regulated by independent bodies, which have specified independent systems for handling emergency calls. While wired networks have continued to rely on circuit-switched technologies developed over the past thirty years, the state's wireless networks, having more of an opportunity for a clean-slate design, have created a system for backhauling calls over IP from the edge of a carrier network to a PSAP, using a dedicated state-wide fiber-optic backbone.

There is a similar heterogeneity in how emergency services are funded across the US. The services themselves (e.g., police, ambulance, fire brigade) are typically funded from taxes collected by the localities they serve. Some services also collect fees from people who call for help, especially ambulance services. PSAPs are funded by a mixture of local, state and federal funds; how much is contributed by each level, however, varies widely from one jurisdiction to another. On the carrier side, telecom providers are allowed to charge users a fee as part of their subscription in order to cover costs related to 9-1-1, but not to charge for individual 9-1-1 calls. In some jurisdictions, these 9-1-1 fees also help fund local PSAPs.

In light of all these different management frameworks, though, there is still some need for a consistent technical framework, for example, to allow vendors to offer a consistent set of equipment nationwide (and for PSAPs to benefit from reduced prices due to economies of scale). The national coordination organization for emergency services is called the National Emergency Number Association, or NENA.

NENA is a non-profit corporation, formed in 1982 to raise awareness of 9-1-1 and promote national deployment of consistent emergency services. Today, it has over 7,000 members representing PSAPs, government agencies, equipment vendors, and others, supporting a variety of activities related to 9-1-1, including:

- development of standards for emergency services architectures;
- education and certification programs for emergency managers;
- government relations, policy, and legislation development;
- operational coordination of emergency services.

In particular, NENA has had a strong focus on IP-based emergency services through its "Next Generation 9-1-1" or NG9-1-1 initiative. The main output of NG9-1-1 is a set of architectures describing the transition between the current emergency calling system and an eventual pure-IP system. The main transition step is the "i2" architecture, in which IP geolocation technologies are bridged into existing location databases and SIP calls are gatewayed to the PSTN to reach PSAPs. The "i3" architecture is the end goal, in which end devices and access networks use the ECRIT architecture (described in Chapter 3) to deliver calls to an Emergency Services IP network (ESInet). ESInets then allow PSAPs to receive calls, interact with other PSAPs and emergency responders, and obtain other auxiliary services.

NENA also has a Government Affairs director and Regulatory/Legislative committee, who have been working on recommending policy changes at the state and federal level to facilitate NG9-1-1 deployment. For example, the NENA government affairs website, at `http://www.nena.org/government-affairs`, provides model legislation on topics related to 9-1-1. NENA has also published the *Next Generation 9-1-1 Transition Policy Implementation Handbook* (2010) for state and federal policy-makers, which provides government officials with advice on how to create policies to support NG9-1-1 in their jurisdictions. The policy handbook covers a wide range of issues, ranging from the funding of various components of the NG9-1-1 system to tariffing and liability issues.

In summary, there are several organizations (more than 50!) responsible for regulating, funding, and coordinating emergency calling in the US:

- FCC, US Congress – Establishing national standards and providing grants.
- State and local governments – Establishing state-wide standards and operating PSAPs.
- NENA – Coordinating nation-wide technical and operational standards.

5.3.2 9-1-1 History

Emergency calling services were first established in the US by the US Federal Communications Commission (FCC) and AT&T in the late 1960s. The FCC is the US national regulator for telecommunications, created by the Communications Act of 1934 (updated in 1996) for the purpose of "regulating interstate and foreign commerce in communication by wire and radio". At the time, AT&T was the national monopoly telephone company. 9-1-1 was born when the FCC asked AT&T to establish a universal emergency number.

The choice of the specific dial string "9-1-1" is helpful for a few reasons. Because it is a short sequence of numbers that can be used to reach all emergency services, it is easy for callers to remember. Because the digits involved are at the extremities of a touch-tone key pad or rotary dial, callers can find the required numbers easily. More technically, as telephone numbers were defined at the time, the sequence "9-1-1" would never appear at the beginning of any other "normal" phone number, avoiding mistakes in call routing.

More recently, though, the potential for accidents in 9-1-1 dialing has increased due to the way many business phone systems are configured: In many corporate phone systems, callers must first dial a "9" before dialing an outside number. This system leads to two problems: First, it means that callers must actually dial "9-9-1-1" in order to place an emergency call, since otherwise the system would simply read the "9" as a request for an outside line, then dial the meaningless string "1-1" on the outside line. Some systems handle 9-1-1 as a special case, but this leads to a second problem: Long-distance calls in the US are initiated by dialing "1" before the number, so if a caller in a corporate network wants to place a long-distance call on an outside line, he must dial "9", then "1", then the number. If the caller accidentally dials this "1" twice, then he will have inadvertently dialed "9-1-1" and placed an emergency call.

Nonetheless, the 9-1-1 system for emergency calling has been very successful. As of this writing, the national 9-1-1 system is comprised of over 6,000 PSAPs, which handle more than 240 million calls every year. Some 99% of the US population have access to some level of emergency calling service. Full statistics on 9-1-1 in the US, collected from the states by the FCC and summarized by NENA, can be found at `http://www.nena.org/911-statistics`.

As telecommunications networks have evolved, the 9-1-1 service has evolved with them. The first 9-1-1 services simply ensured that the caller reached an operator who was prepared to help with an emergency. For instance, the first 9-1-1 system (in Haleyville, Alabama) simply connected the caller to the local police station. Such basic 9-1-1 services were deployed through the late 1960s and 1970s. In the late 1970s and 1980s, as automated data systems began to be widely available, there was a migration toward "Enhanced 9-1-1" or "E9-1-1", in which calls are directed to different places depending on location of the calling line. Moreover, the caller location information used to route the call is also conveyed to the call taker that responds to the call, all in an automated fashion.

The integration of location information was an important innovation for emergency calling in the US, because the management of the emergency calling system is so decentralized. Given that there are around 6,000 PSAPs in the US, the average PSAP covers an area of less than 600 square miles (around 1,500 square kilometers). So routing an emergency call to the correct PSAP requires location information that is quite precise; if the average jurisdiction covering 600 square miles were a circle, it would have a radius of only 14 miles (22km). Simply routing all emergency calls from a switching center that covers a city or county to the same destination would result in a large number of calls arriving at a PSAP in a different jurisdiction.

Thus, the advent of cellular telephones caused significant problems for emergency calling, and led to "Wireless Enhanced 9-1-1", an extension of E9-1-1 in which cellular providers dynamically update location records so that the PSAP that receives the call has up-to-date information about the caller's location. The FCC regulation defining wireless E9-1-1 provides two phases of deployment: In Phase I, the cellular provider needs to provide the location of the cell tower to which the caller is connected, while in Phase II, the provider needs to provide the actual location of the caller to within 300 meters.

Currently, the E9-1-1 system is evolving toward a "next-generation" 9-1-1 system, following the NENA "NG9-1-1" guidelines for supporting emergency services over IP. As mentioned above, this effort is proceeding in stages, much like the effort to enable E9-1-1 for wireless; in this case, the different steps in the transition are defined by the NENA "i" architectures, primarily "i2" and "i3" (*Interim VoIP Architecture for Enhanced 9-1-1 Services (i2)*, 2005; *NENA Functional and Interface Standards for Next Generation 9-1-1 Version 1.0 (i3)*, 2007). The "Interim VoIP Architecture for Enhanced 9-1-1 Services" (i2) defines a system to allow VoIP callers to access the emergency calling facilities of the PSTN, as illustrated in Figure 5.2 (*Interim VoIP Architecture for Enhanced 9-1-1 Services (i2)*, 2005). The two critical elements in the i2 system are both gateways between the IP network and the PSTN: The Emergency Services Gateway (ESGW) translates between VoIP protocols for signaling and media (e.g., SIP and RTP) and their PSTN equivalents, and the VoIP Positioning Center (VPC) gives the traditional ALI servers a way to access location information for IP endpoints. At a high level, i2-enabled VoIP providers can provide 9-1-1 access by

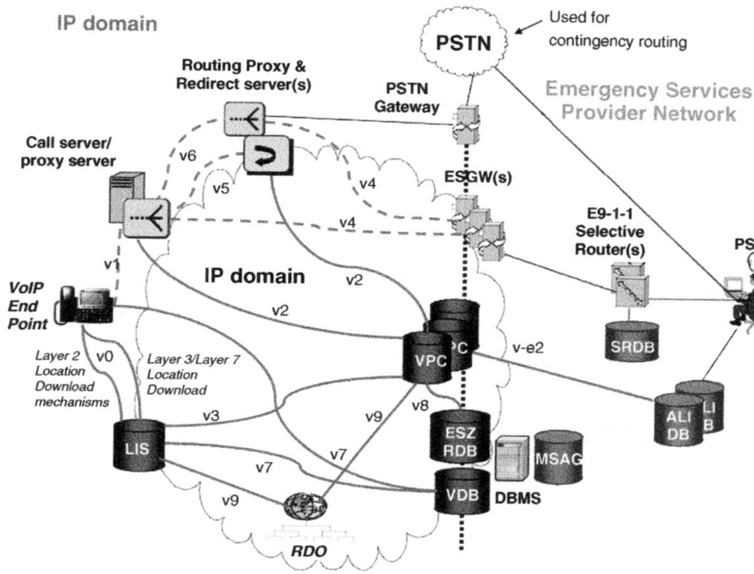

Figure 5.2 The NENA i2 architecture. Reproduced by permission of © 2005 National Emergency Number Association (NENA).

routing calls to the proper ESGW (which will relay them to a PSAP over the PSTN) and by providing access to location information (e.g., subscriber information) through a VPC.

If i2 can be viewed as providing IP endpoints with access to PSTN-based emergency services, then i3 describes a system for all-IP emergency services and provides PSTN endpoints with access to them. From the perspective of callers and networks, the i3 and ECRIT architectures are essentially the same. This congruence is deliberate; much of the work on the ECRIT architecture was contributed by i3 participants, and vice versa. The subject where i3 provides significantly more information, though, is the question of how PSAPs connect to the Internet and receive calls. Naïvely, PSAPs need only be reachable over IP and SIP, but such simple connectivity raises significant security concerns (discussed in Chapter 7). To address these concerns, and to provide additional services to PSAPs, i3 defines the concept of an Emergency Services IP network (ESInet) which provides a secure

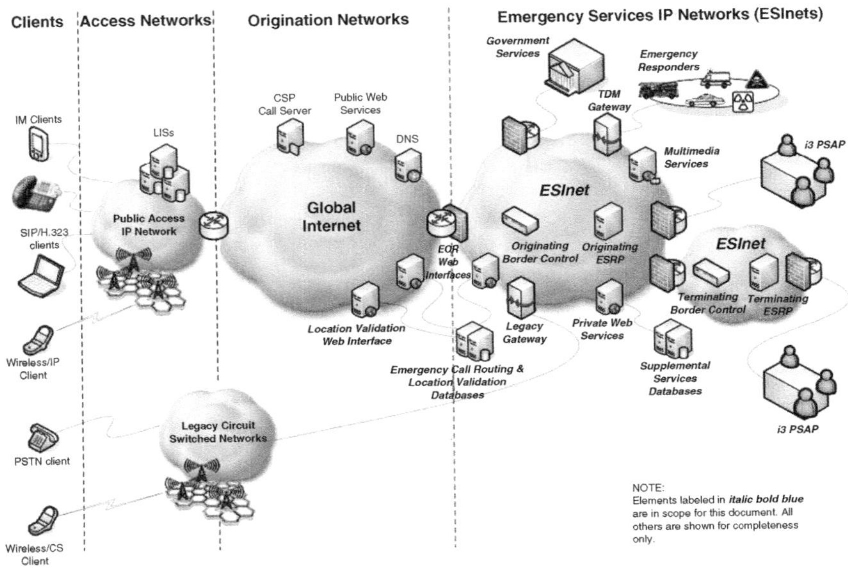

Figure 5.3 The NENA i3 architecture. Reproduced by permission of © 2006 National Emergency Number Association (NENA).

network through PSAPs which can safely connect to the Internet, and through which value-added services (e.g., geocoding) can be made available to PSAPs. The overall i3 system is illustrated in Figure 5.3.

5.3.3 Automatic Location Information

Location information in E9-1-1 is provided by means of an "Automatic Location Information" or ALI databases. For fixed lines, these databases contain records that map telephone numbers to addresses, so that when a PSAP gets a call, it can use the calling number associated with the call ("Automatic Number Information" or ANI) to look up the caller's location. To support mobile phones, these databases were augmented so that they could be dynamically updated by a cellular provider in order to provide current caller location information.

Extending this system to support VoIP 9-1-1 has been a challenge, mainly due to the fact that in VoIP there is not necessarily a relationship between the voice application and the underlying physical network.

Such a relationship is presumed in the structure of the ALI database, which joins an application-layer identifier (i.e., a phone number) with a physical location. In VoIP, neither the voice application nor the physical network has enough information to provide such a record – the voice application knows the phone number but not the location, and vice versa for the network.

Current systems for VoIP emergency calling rely on users registering their location in a web form, so that their VoIP service provider can enter this information with the user's phone number into the appropriate ALI database. This solution clearly has problems from technical and operational perspectives: User-entered location information can be inaccurate or out-of-date, and VoIP providers have to figure out which ALI databases should be provisioned (there is no automated mechanism to find the ALI database for a given jurisdiction).

Similar challenges arise for organizations that operate their own Multi-Line Telephone System (MLTS) or Private Branch Exchange (PBX). Such systems are frequently used in corporate telephone networks, so that the company can provide telephone access for many users via a single trunk provided by the local telco. However, because it is the local telco that supplies information to the ALI database, this arrangement frequently results in all of the lines behind the PBX being assigned the same ALI location, namely the location of the trunk or PBX. This location can be far away from the caller's location – in a different building in a corporate complex or even around the world for a remote worker (indeed, many PBX systems are now IP-based). This problem is mostly solved today by requiring the organization operating the PBX to add records for the lines served by the PBX to the relevant ALI database, often at significant cost.

The system of ALI databases has proven to be a valuable resource for 9-1-1, providing PSAPs with fast, automated access to location information. However, as communications networks are put together in more flexible ways – as voice is handled separately from network access, and as non-telco entities are operating networks – the limitations of the ALI system reduce its applicability to emergency calls over emerging communications media. So the location architecture envisioned for the Internet (described in detail in subsequent chapters) can provide significant benefit both by allowing more flexible access to existing types of location records (e.g., by supporting dynamic discovery of location databases) and by providing extensibility to support new types of communications systems.

5.4 The European Union

The European Union has a somewhat similar situation to the US with regard to emergency calling. In both communities, there is some top-level coordination, but for the most part, decisions are made autonomously by the constituent states and lower-level organizations. In Europe, however, the degree of decentralization and heterogeneity of systems is much higher, in part due to the greater autonomy of European nations (compared to American states). There has also been little effort at Europe-wide coordination of emergency services until relatively recently – indeed, many European emergency calling systems pre-date the EU itself – so by the time efforts at coordination began, most member states had already developed their own emergency calling systems, often in very different ways.

That said, there has begun to be some coordination at both technical and regulatory levels. From a legal and regulatory perspective, the European Parliament has created EU-wide laws governing emergency access – in particular, establishing the universal emergency number 112 – and set an overall regulatory framework. A regulatory coordination group known as BEREC (the Body of European Regulators for Electronic Communications) has been formed to coordinate regulation among member states. Technical coordination is being done in industry groups, including the European Emergency Number Association (EENA) as well as national bodies like NICC in the UK and AK-TK in Austria).

The Body of European Regulators for Electronic Communications (BEREC, the successor to the European Regulators Group) was established by the European Commission in order to foster and improve communication and cooperation among the several national regulators and the Commission. The body also has a goal of unifying the various regulatory frameworks of member states. Recommendations will thus surely affect the situation in many of the member states of the European Union. Working plans as well as further information are available at <http://berec.europa.eu/>. BEREC consists of the heads of the 27 national regulatory authorities ensuring consistency of regulation, exchanging expertise, best practices and transmitting information.

Currently, a reform of the "Regulatory Framework" is in progress, which will also affect emergency services. It is possible that this effort will result in more stringent requirements for mobile networks to deliver location information, and for VoIP operators to do the same.

Beyond the BEREC, at the European level, there is also the "Expert Group on Emergency Access", which is working out general emergency calling requirements for EU member states, in particular with respect to the interface between public networks and the emergency calling system (see, e.g., the "Operational Needs Document" (*Operational Needs for Access to Emergency Services*, 2007)).

The European Emergency Number Association is increasingly acting as a central coordination point for non-governmental participants in emergency calling, especially for technical coordination. Initially formed as a lobbying group to promote legislative and regulatory actions related to 112, EENA recently formed an "NG112 technical committee" to help promote emergency service standards and interoperability among EU states, much as NENA has done in the US. The NG112 group is comprised of volunteers from a variety of organizations, including equipment vendors and service providers that have been supporting national emergency services deployments and standards efforts, as well as other standards organizations such as the IETF, 3GPP, and NENA. The EENA technical committee is currently in the process of evaluating the NENA i3 architecture, comparing it to requirements of European emergency services networks, with the goal of creating an internationally interoperable framework.

5.5 Japan

The emergency calling system in Japan represents a mid-point between a single, unitary system (as in Austria) and a distributed, heterogeneous system (as in the US or the EU). Emergency calling requirements are set at the national level, and there is a high degree of national coordination on implementation, but individual PSAPs are still run in a more decentralized fashion. The Japanese emergency calling system also has a few unique features that have created unusual requirements for the transition to VoIP.

5.5.1 Regulatory Framework

Emergency services in Japan are administered by the national Ministry for Internal Affairs and Communication (MIC), as part of its general duty to regulate the telecommunications industry established by the Telecommunications Business Act (*Telecommunications Business Act (Japan)*, 2007). The Telecommunications Business Act was first

Table 5.4 Emergency dial strings for Japan

Number	Service
110	Police
118	Coastguard / Maritime safety
119	Fire / Ambulance

enacted in 1984, and it has been revised a number of times, most recently in 2007. The specific regulations governing telecommunications are laid out in regulations issued under this law, primarily in the form of "Ministerial Ordinances". For example, Order number 25 of 1985 of the Ministry of Posts and Telecommunications (the predecessor to the MIC) defines the overall regulations for telecommunications (e.g., classes of service) called for by the Telecommunications Business Act, and defines enforcement provisions for these rules (*Regulations for Enforcement of the Telecommunications Business Act (Ministerial Ordinance of MPT No. 25 of 1985)*, 2007).

There are three different emergency numbers in Japan for three emergency services (see Table 5.4). 110 is used to reach the police, 118 for the Coastguard and maritime safety, and 119 to reach fire and ambulance services.

Operation of PSAPs is delegated to the individual services, and each divides up this responsibility a little bit differently. The police maintain many small PSAPs in each region throughout the country, while the fire, ambulance, and maritime services are run in a more central manner. The changes in carrier networks that are necessary to support emergency services are coordinated among the carriers and the government via the MIC.

Japanese regulation defines a well-established division of IP telephony services into two classes: roughly speaking, those that provide service equivalent to the PSTN and those that do not. These two classes are distinguished by being assigned telephone numbers under different area codes, PSTN-equivalent services are numbered with the standard scheme (commonly referred to as "0AB~J" numbers), and other services are numbered with the special area code "050".

Criteria for an IP telephony provider to obtain a 0AB~J designation are fairly stringent: The provider's network provides sufficiently low call failure rates and end-to-end round-trip times, and must score sufficiently high on the ITU-T G.107 voice quality rating. Moreover, endpoints must not be nomadic, in the sense that they cannot be removed from one premises and re-installed in another without the provider's knowledge (a common feature of many VoIP services). 050 services, of course, do not need to meet these criteria; they only need a point of interconnection to the PSTN. In particular, services that incorporate soft phones are necessarily 050 services, since software-based phones are inherently nomadic.

Of course, in order to be equivalent to the PSTN, 0AB~J providers are required to provide standard emergency services to subscribers. 050 providers do not have an obligation to provide emergency services, but there is a system for them to do so if they so desire (e.g., if they are in the process of upgrading to meet the 0AB~J criteria).

Figure 5.4 shows the trend in the allocation of 0AB~J and 050 numbers over the past few years. These data (courtesy of MIC) clearly show an increasing prevalence of 0AB~J services over 050 services. There are a couple reasons for this trend. First, Japan has seen a large growth in fiber-to-the-home (FTTH) deployments in the past few years. As in much of the world, FTTH deployment gives carriers an opportunity to replace traditional PSTN connectivity with FTTH-based IP telephony, in a way that is transparent to the user (the FTTH endpoint translates between the normal PSTN protocols within the customer premises and IP protocols for the network). Thus, much of this growth is being driven by new deployments of PSTN-replacement IP phones. (This also explains why growth in 0AB~J devices is not at the expense of 050 devices.) In addition, there is a perception in the marketplace that 050 services are cheaper and of lower quality, which is an important consideration, for example, for businesses using these services.

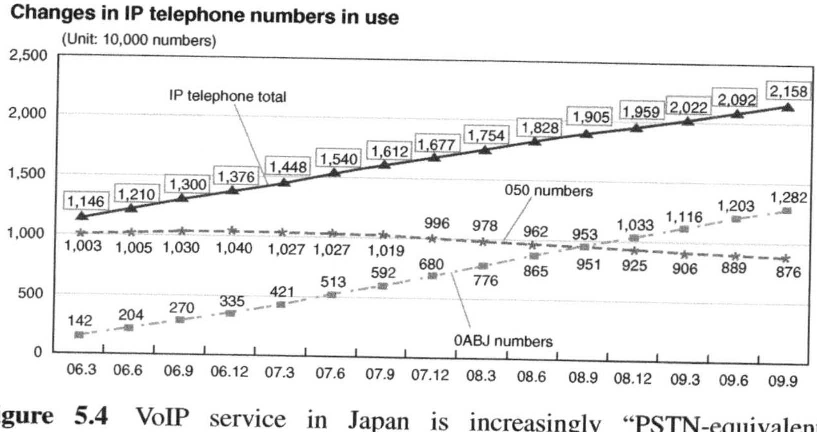

Figure 5.4 VoIP service in Japan is increasingly "PSTN-equivalent" (0AB~J). *Announcement of Quarterly Data Concerning Subscriber Numbers and Market Share in Telecommunications Services (Second Quarter FY2009)* (2010).

Of course, VoIP services that do not use phone numbers cannot interconnect with the PSTN, and do not seem to be subject to regulation. In the jargon, these services are called "Internet telephone" services, as opposed to "IP telephone" services.

5.5.2 *Call Handling*

As illustrated in Figure 5.5, the basic call flow for IP emergency calls in Japan is similar to that in other parts of the world right now. All PSAPs are connected to the PSTN, so the VoIP operator has to recognize when emergency calls are being placed and route them through an appropriate gateway to the PSTN. Emergency calls are expected to be given higher priority than normal calls, in both the IP and PSTN domains. VoIP operators are also required to operate a Location Server that makes location accessible to PSAPs, as discussed in more detail below. In these regards, the Japanese system is analogous to the NENA i2 architecture and other national systems that make PSTN emergency calling functions available via national VoIP networks.

The general system, in which the VoIP provider operates a gateway and a Location Server, and properly routes and prioritizes emergency

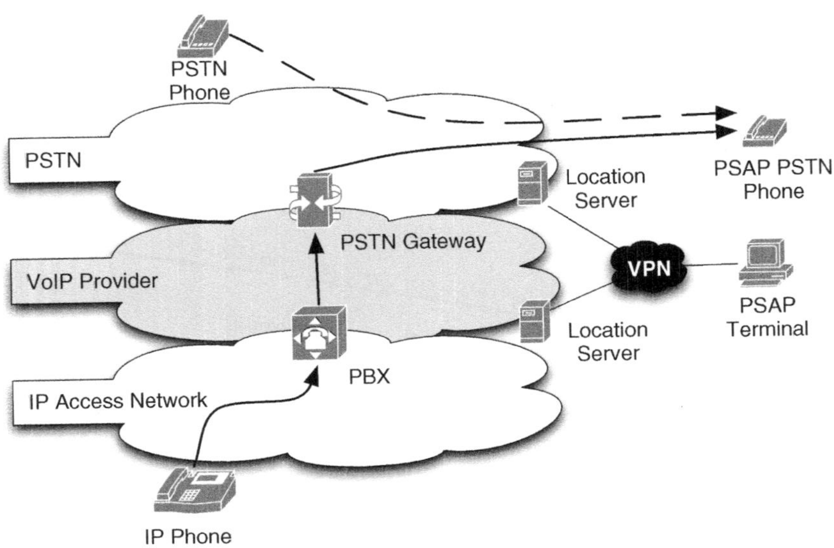

Figure 5.5 High-level architecture for how emergency calls are delivered to PSTN-connected emergency services in Japan (and elsewhere).

calls, is the same whether the VoIP provider is an 0AB~J provider or an 050 provider. Only 0AB~J providers are required to implement such emergency services.

The Japanese system has one unusual feature, however, which has inspired debate in the Internet community: Callers are not allowed to disconnect a call. In the PSTN, this feature is relatively straightforward to implement. The control plane simply maintains the circuit between the caller's device and the PSAP until the call taker in the PSAP disconnects it. The caller's device will remain connected to the PSAP, unable to place another call, and media will continue to flow between the device and the PSAP. This system has been implemented in Japan and a few other jurisdictions (e.g., parts of Canada).

For VoIP systems, the situation is considerably less clear. First, there is of course no notion of a "circuit" in VoIP, only a logical "session" that the two endpoints to the call can join or leave at will. Second, in many VoIP systems, the endpoints are completely autonomous from the rest of the system, in that control elements can send messages to an endpoint, but have no way to compel it to take action on that message. Debate about how to implement this so-called "premature disconnect prevention" feature for VoIP in general is ongoing in the IETF ECRIT working group and other groups. Because the notion of network control is so ill-defined for VoIP in general, there have been proposals that all that can be said in general is that the user interface of the calling device should prevent (or at least deter) the user from disconnecting an emergency call (in effect, ceding control over disconnection to the network).

In the Japanese context for VoIP, there is an intermediate solution in place, which combines aspects of the PSTN and VoIP solutions. Since there is a PSTN connection between the service provider's PSTN gateway and the PSAP, the normal PSTN premature disconnect feature is enabled on that connection. On the VoIP side of the gateway, if the caller disconnects, then the gateway attempts to re-establish the session by sending a SIP INVITE message to the caller (i.e., ringing the caller's phone) every few seconds, until either the caller picks up the phone or the PSAP disconnects.

5.5.3 *Location Information and Privacy*

Automatic location delivery is a core feature of the Japanese emergency calling system. Since at least 2007, both fixed and mobile operators have been required to deliver location information for the caller at the start of an emergency call, through an automated Emergency Location Reporting System (*Emergency Location Reporting System – Providing Location Information in Emergencies Using Mobile Telephones*, 2007). In this system, each operator creates a Location Server and connects it to VPN, Location Servers can send updates to PSAPs and PSAPs can query Location Servers. These transactions are conducted using standard XML protocols over HTTP. For example, Location Servers for mobile operators use the OMA-standard Mobile Location Protocol.

Again, in the spirit of PSTN equivalence, 0AB~J VoIP operators are required to operate such a Location Server and connect it to the location VPN. 050 operators are required to do so if they choose to implement emergency services.

The sources of the location information that these servers provide vary according to the type of network being served. For fixed networks and 0AB~J VoIP networks, where the nature of the network prevents endpoints from being mobile, location information is stable (basically a service address for the subscriber), and can be statically provisioned into the Location Server. For mobile networks, location information is typically based on GPS or assisted-GPS when possible; otherwise, the location of the serving base station is returned.

The situation is more difficult for 050 VoIP networks, where endpoints can be nomadic, but the VoIP provider may not necessarily be able to dynamically locate them. Because no standard system for IP location is in place (which would allow such a dynamic location), the location values provided by 050 networks are typically entered manually by the user or based on subscription information – with all the risks of inaccuracy and lack of timeliness that such methods entail. For nomadic services, however, this system is essentially the best that can be done without an automated system for general IP geolocation, and it is a common pattern in several countries. (E.g., this is typically how VoIP providers meet E9-1-1 requirements in the US.) A central goal of the ECRIT architecture (and the GEOPRIV system it relies on) is to address this gap, as described in Chapter 4.

The Japanese system, however, adds an unique additional feature on top of all these automated location systems, namely location privacy: By dialing "184" before an emergency number, a user can prevent location information from being sent along with his emergency call, and if the caller dials "186", calling number information is suppressed as well. So even though location is provided to PSAPs automatically by the voice network, the networks are required by law to withhold user location information in these circumstances. There is, however, an override: If the call taker that answers a call decides that lives are at risk that could be saved if location information were available, then he can force the caller's location to be revealed.

Like the "premature disconnect" feature discussed above, it is not immediately clear how this location suppression feature would be implemented in the general VoIP context. The problem in general is that a VoIP operator can be in a completely different jurisdiction from the destination PSAP, and VoIP operators need location information in order to route calls to the correct PSAP. Thus, the VoIP operator would need some signal to indicate that it should strip location from the call before delivering it to the PSAP; none of the current ECRIT protocols provide such a semantic. The situation is further complicated by the fact that queries for location are likely to be answered by the caller's ISP (as opposed to the VoIP provider). Because of the separation between the IP layer and the voice layer, the ISP may have no idea what number the caller dialed. In the end, this feature may need to be implemented in the user interface of PSAP equipment, by not showing a call taker location information for calls to designated numbers.

5.6 Summary

In this section, we have examined the emergency services systems in a few different countries. They vary quite a bit in terms of how emergency calling systems are regulated and managed, with centralized systems in Austria and Japan and coordinated but heterogeneous systems in the US and the EU (considered as a whole). With location information in particular, there is significant variability in deployed capabilities, ranging from a manual exchange of faxes in Austria, to circuit-based ALI systems in the US, to modern, IP-based systems in Japan.

Nonetheless, there are a few areas of commonality, especially when it comes to emerging regulations for Voice-over-IP services. In many cases, VoIP services that interconnect with the PSTN are the first to be subject to emergency services regulation, in part because they offer users a service that is frequently indistinguishable from PSTN service, but also in part because their need for PSTN numbers forces an interface between the VoIP provider and the national government controlling the phone numbers. In many cases, rules surrounding number allocation force the VoIP provider to have a legal presence in a country, providing an explicit target for regulation. Given this situation, there are already some systems for addressing the particular case of "interconnected"

VoIP – some emerging, as in the US, and some more mature, as in Japan. Even these solutions, however, only deal really well with cases that are essentially equivalent to PSTN networks, in which the network service and the application service are provided by the same entity.

For "nomadic" interconnected VoIP services, and general VoIP services, there is still a need for a more thorough solution. Current solutions – in the US and Japan as described above, but also elsewhere – frequently rely on manually-entered location for call routing, and manually-configured gateways to emergency services. It is solving these problems, the establishment of automatic location for general IP networks and dynamic discovery of emergency services, that constitute the "context resolution" function that is the core of the ECRIT architecture.

References

7 Millionen Argumente fuer Mobilfunk-Infrastruktur (2010). Forum Mobilkommunikation.

Announcement of Quarterly Data Concerning Subscriber Numbers and Market Share in Telecommunications Services (Second Quarter FY2009) (2010). MIC Communications News.

Emergency Location Reporting System – Providing Location Information in Emergencies Using Mobile Telephones (2007). MIC Communications News.

Handy Am Berg (2006). Telekom Austria.

Interim VoIP Architecture for Enhanced 9-1-1 Services (i2) (2005). National Emergency Numbering Association (US).

IP-Enabled Services E911 Requirements for IP-Enabled Service Providers (2005). FCC.

Leitweglenkung und die Standortidentifikation der Notrufe (2005). Bundesamt für Kommunikation (BAKOM).

NENA Functional and Interface Standards for Next Generation 9-1-1 Version 1.0 (i3) (2007). National Emergency Numbering Association (US).

Next Generation 9-1-1 Transition Policy Implementation Handbook (2010). National Emergency Numbering Association (US).

Operational Needs for Access to Emergency Services (2007). European Commission, Expert Group on Emergency Access.

Qualitätskriterien für den Universaldienst gemäß Universaldienstordnung (2007). Telekom Austria.

Regulations for Enforcement of the Telecommunications Business Act (Ministerial Ordinance of MPT No. 25 of 1985) (2007).

Richtlinien für Anbieter von VoIP Diensten (2005). RTR GmbH.

Rosen B and Polk J (2010) Best Current Practice for Communications Services in support of Emergency Calling. Internet Draft (work in progress) draft-ietf-ecrit-phonebcp.

Stastny R and Merka M (2006) "Next Generation emergency services: die Zukunft der Notrufe". *e & i Elektrotechnik und Informationstechnik* pp. 323–332.

Telecommunications Business Act (Japan) (2007).

Zusammenschaltungsvertrag (2008). Telekom Austria.

6

VoIP Emergency Calling in Practice

In the preceding chapters, we discussed the ECRIT emergency calling architecture and the process of emergency calling in the abstract. In order to make some of these concepts more concrete, this chapter gives an overview of some tools that have already been created to implement emergency calling, as well as some recommendations on how to deploy these services. These tools and recommendations can be used to develop prototype emergency calling systems so that engineers can gain experience with the new technologies involved or test out production systems as they are being developed. Certainly, at this stage these tools cannot be used for placing real emergency calls – these are prototypes for experimenting only. Hence, when referring to emergency calls and PSAPs in the following sections, we mean to include only test calls in an experimental set-up.

In the following sections, we will discuss several software tools (or extensions to existing tools) that help with different parts of the emergency calling process, including providing endpoints to location, guiding to the right PSAP, and helping them make an emergency call.

With the help of these tools, one can easily test out VoIP emergency calling in a lab setting. Since the relevant standards are still evolving a little, some tools are slightly out of line with current standards. However, most of the tools discussed here are still under active development, and expected to ultimately align with standards. In any case, the tools and exercises in this chapter can be useful to get a basic understanding of how VoIP emergency calling works.

6.1 Software

In this section, we'll describe several open source software projects that can be useful for implementing emergency calling for VoIP, including

VoIP Emergency Calling: Foundations and Practice Karl Heinz Wolf and Richard Barnes

some new projects as well as some extensions to existing software packages. These tools collectively cover most aspects of the emergency calling process:

- Location Configuration:
  - Open-source HELD servers and clients.
  - DHCP encoders and decoders.
  - Wireshark extensions for location protocols.
  - OpenLLDP.
  - HELD support in Firefox.
- Location-to-Service Mapping:
  - Columbia University LoST client and server.
  - University of Göttingen LoST client and server.
  - Krakow Polytechnic University LoST client and server.
- Emergency call handling:
  - Emergency calling extensions to the Zap! SIP client.
  - Ecritdroid client for Android.
  - EcritXUL extension for Firefox.
  - Location extensions to the Asterisk PBX.
  - Open IMS Core Emergency Services.

A full list of open-source tools for experimenting with ECRIT technologies will be maintained on the companion website for this book, at `http://www.voip-sos.net/tools/`.

If you want to try out only a few specific parts of the ECRIT architecture, you won't need to install all of the tools we discuss here. For instance, to test location configuration via HELD, you would only need to install a HELD server and a HELD client. At the end of this chapter we'll describe a few example set-ups.

6.1.1 HELD Clients and Servers

Because the HELD protocol is carried in XML messages over HTTP, it's relatively straightforward to create software that implements it using the wide variety of HTTP and XML libraries out there. So there are already a few different open-source implementations of HELD, both on the client side and the server side.

Note, however, that these implementations typically don't do much to address the hard part of deploying a location service, that is, actually finding out where the target is in the network. Typically, open-source servers will define a few simple mechanisms to provide location (e.g., statically provisioning location or using an existing IP-to-geo service), but they define an interface that individual networks can use to plug in new positioning modules. So the HELD servers discussed here can be a good starting point for exploring how to do positioning in your own network.

The Open Source Location Information Server (LIS) project (sponsored by Andrew Corporation) has created client and server implementations of the HELD protocol. The main project website is hosted on SourceForge at `http://held-location.sourceforge.net/`. The server is based on PHP, and uses the Postgres database system for storage, so it can run on many standard web servers. For positioning, the server offers two possibilities: Either the server can be supplied with static mappings of IP addresses to location values, or the server can locate the client using "switch-port mapping". (In the latter case, the LIS queries switches over to SNMP for information about which switch and port a client is connected to. The location returned to the client is then the location of the opposite end of the wire connected to that port, e.g., a wall jack in an office.) Other types of positioning can be integrated by creating a PHP module that supports them.

The client provided by the Open Source LIS project is a Java implementation of HELD, with a wrapper to allow it to run as a Windows service. A single instance of the client service can thus do the work of discovering the local Location Server and acquiring location information, then make it available to applications that want location. Applications then access the client via the Windows services mechanism, using a simple XML-based interface.

A team at BBN Technologies have begun work on an "Internet geolocation toolkit" (igtk), which is intended to eventually provide an entire suite of tools for dealing with geolocation on the Internet. The main project site is on SourceForge at `http://igtk.sourceforge.net/`. The current version of the toolkit provides a few different tools, including a HELD client library, as well as a few HELD server implementations, and some location-based applications that use the HELD client library.

The igtk HELD client library is written in C++, with the stated goal of facilitating integration into applications in many languages, or possibly eventually as an operating system service. At the same time, though, it is also a goal for this library to remain platform independent, and it is regularly tested on Windows, Linux, and Mac OS, all in multiple versions. (The library also includes code to parse the DHCP location options.)

In order to demonstrate how the client library can be used with multiple languages, the project includes a few different location-based applications. On the C++ side, there is a basic command-line application that simply prints out the location it gets from the server. The project also includes a wrapper that enables Java applications to use the HELD client library, as well as a framework for building applications that push location updates to location-based services (e.g., the Yahoo! FireEagle service).

igtk includes two different Location Servers, one in Perl and one in Java. The Perl LIS is fairly basic, supporting two basic positioning mechanisms: It can draw on the MaxMind geo-IP databases (via the Perl interface provided by MaxMind), but it also includes a provisioning interface that an operator can use to specify the location of a block of IP addresses, illustrated in Figure 6.1. The Java LIS was created in an effort to facilitate translation among different protocols, and in particular, to make it easy to provide a HELD interface to existing location resources. So the Java LIS is based on a system of components, with "servlet components" on the client-facing side that implement server logic for various protocols (e.g., HELD), "source components" that pull location from databases or act as clients for other protocols,

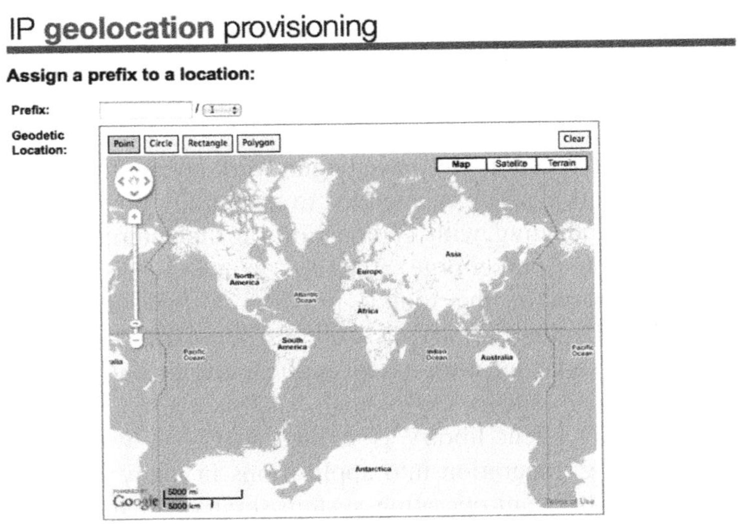

Figure 6.1 Provisioning the igtk Perl LIS.

and "service components" that are set between sources and servlets to add additional features, such as caching or logging.

6.1.2 DHCP Location Encoders and Decoders

In contrast to HELD, which requires special code for clients and servers, most DHCP servers are able to send arbitrary options, and most DHCP clients are able to receive and store them. So most of the work in implementing DHCP-based location configuration is in encoding the right location values into the binary form used by DHCP, then decoding them on the client side.

Using DHCP can be very straightforward if a DHCP server covers a single geographic area – in particular, if all the clients of the DHCP server are served by the same PSAP. In that case, the operator of the DHCP server can simply supply a static location value that describes the whole network. For instance, an enterprise network that covers an office building could simply be configured with the location of the building. A Cisco DHCP server could be configured to tell all clients they are in the United States, area code 20001 (part of Washington, DC) by adding the following line to the server's configuration:

```
option 99 hex 02:55:53:18:05:32:30:30:30:31
```

If the DHCP server needs to provide different location values for different subnets or clients, there is slightly more work involved than generating a single DHCP option, but depending on the DHCP server and the level of granularity desired, this can often still be done using configuration files rather than new software.

Note, however, that the granularity of location provided by DHCP is directly related to the utility of location for responding to an emergency. Location that is accurate at the post code or building level will enable callers to reach a PSAP, but may not provide the PSAP with enough information to direct responders to the site of an emergency (e.g., an office within a large building). So even though it is more straightforward to provision coarse-grained location, a more precise location is, as always, more desirable.

RFC 3825 describes how coordinates can be encoded in DHCP for the purposes of location configuration. In this system, coordinates and uncertainty information are encoded in a packed binary representation that is not human-readable. nic.at has developed an RFC 3825 encoder/decoder that can translate between this binary representation and a human-readable one. In addition, since LLDP-MED uses the same location format as DHCP, this tool can also be used to handle location information from LLDP-MED. The encoder/decoder tool can be downloaded from `http://www.voip-sos.net/tools/rfc3825/`.

In order to use DHCP location configuration to provide endpoints with civic address information, it is necessary to encode the address according to the format described in RFC 4776. This encoding can be performed using the online service available at `http://www.voip-sos.net/tools/rfc4776/`. This service already incorporates the clarifications specified in RFC 5139. Decoding of this DHCP option isn't provided by the online service, but Wireshark will decode addresses observed in DHCP traffic.

For a slightly more advanced interface, there is also an AJAX-based encoder on the website, at `http://www.voip-sos.net/tools/dhcloc/`, shown in Figure 6.2. This encoder provides a Google Maps interface for encoding geodetic location, and a more flexible interface for specifying civic location (e.g., you can specify values in more than one language). The AJAX encoder also generates configuration lines for Cisco and ISC DHCP servers that can be copied and pasted into DHCP server configuration files to supply the encoded value as a static option.

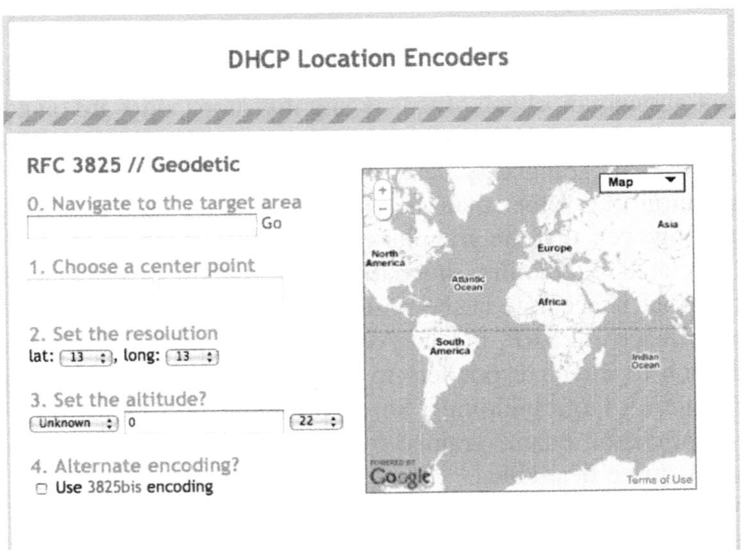

Figure 6.2 The AJAX location encoder for DHCP.

6.1.3 *Wireshark for DHCP Location*

Wireshark is a very well-known and popular tool for collecting and analyzing network traffic. General information on Wireshark is available at `http://www.wireshark.org/`.

One of the useful features of Wireshark is that it "decodes" packets in order to display binary packet payloads in a way that's easy for a human to understand. nic.at has developed extensions to Wireshark to allow it to decode both civic and geodetic DHCP options for location information. The current stable version of Wireshark has already incorporated the extension to handle location information in the form of an address (i.e., civic location (Schulzrinne, 2006)), as shown in Figure 6.3 as well as the extension to handle geolocation in coordinate form (i.e., geodetic location (Polk et al., 2004)).

6.1.4 *OpenLLDP*

The OpenLLDP project has created an open-source implementation of the LLDP protocol. The main project page is at `http://openlldp.sourceforge.net/`. This implementation of LLDP is planned to support several platforms, including PCs (Linux

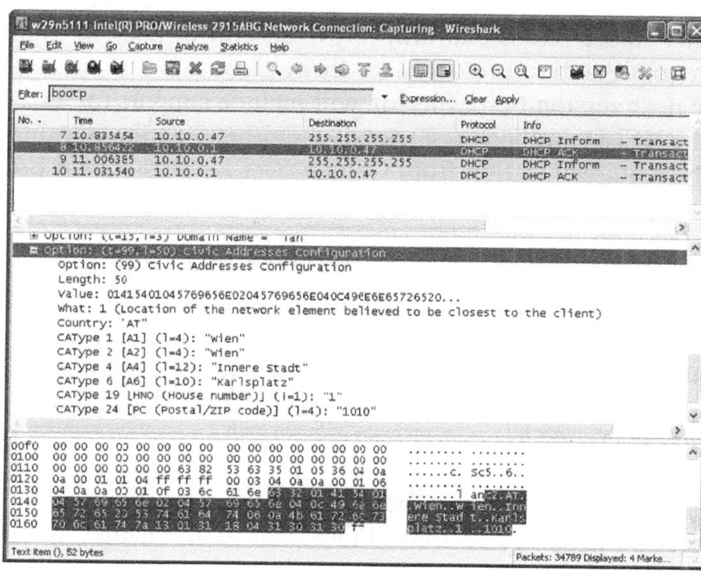

Figure 6.3 Wireshark decodes DHCP location information. Wireshark and the "fin" logo are registered trademarks of the Wireshark Foundation.

supported and Windows and Mac OS X under development) and, for example, switches that can run the OpenWRT firmware. In connection with emergency calls, LLDP is the basis for the LLDP-MED protocol for delivering location information to clients at layer 2 (an optional part of the ECRIT model). To support the use of OpenLLDP for emergency calling, nic.at contributed a patch to this project that enables the LLDP-MED extensions, in particular for location. Until a new OpenLLDP version is released, you will have to download the latest OpenLLDP source from the OpenLLDP website and compile it to enable the LLDP-MED location extension.

6.1.5 HELD Support in Firefox

Several major browsers currently support the W3C Geolocation API (*Geolocation API Specification*, 2010), including Firefox, Google Chrome, Safari, and Opera (in beta as of this writing). The Geolocation API allows the browser to provide location to web pages via a standard Javascript interface. The standardization of this interface has raised the question of how browsers should find location information. Browsers on some platforms, especially mobile platforms, are able to use an underlying location API; this is how the Geolocation API is supported on the browser for the Apple iOS operating system and on Android. But other more cross-platform browsers have had to implement location configuration protocols for themselves.

In the next version (4.0), Firefox will include support for a basic profile of the HELD protocol as a way of acquiring location information (it currently acquires location from Google over a non-standard protocol). HELD support will be included as an optional component, which can be enabled with the following steps:

1. Go to the advanced configuration screen by entering “about:config” in the location bar.
2. Change the value of the setting “geo.wifi.protocol” to “1”.
3. Change the value of the setting “geo.wifi.uri” to point to a HELD server.

Once HELD support has been activated, there should be no user-visible changes; the only thing that will change is the way that the browser acquires location information to provide to web pages via the Geolocation API. You can observe this behavior by going to any web site that uses the Geolocation API, such as http://www.voip-sos.net/tools/web-geo/. If you watch the network traffic in Wireshark, you should see a HELD request to the server you configured when the Geolocation API is invoked. So while this extension doesn't bring much in the way of new features, it can be a handy tool for testing a HELD server.

6.1.6 LoST Implementations

There are a few different open-source implementations of the LoST protocol developed by different academic groups around the world. Each implementation provides a client implementation and a server implementation, and they seem to interoperate with each other, at least at a basic level.

Dr. Henning Schulzrinne's group at Columbia University has been at the forefront of ECRIT development. As part of their work on ECRIT, they have developed an open source client and server implementation of the LoST protocol for using location information to find information about emergency services and PSAPs. Both the client and server are written in Java, and the server uses the geolocation extensions to the Postgres database to store mapping information. On the project website, http://honamsun.cs.columbia.edu/index.html, one can find complete information on the status of the project, as well as source code for the client and server.

In addition, the Columbia group provides a web client that can send LoST queries to a Columbia-operated LoST server that is accessible to the public.

- Web client: http://honamsun.cs.columbia.edu/client.html.
- LoST server: http://ng911serv.irt.cs.columbia.edu:8080/lost/LoSTServlet.

This server is of course provisioned only with example data for a few regions of the US (mainly New York City and northern New Jersey); the examples on the website show locations for which the server has information. The PSAP addresses provided by the server are only examples and cannot receive emergency calls. Even though this server only contains example information, it can still be useful for testing out LoST client implementations, such as the one in the Zap! emergency calling client.

In addition to the Columbia implementation, a team at the University of Göttingen has developed a LoST server based on PHP, using the

MySQL database as back-end storage. This server can be accessed using the Columbia client, or with one of the two clients that the Göttingen team developed, one in Javascript, and one as a PHP web service. The main page for the project is hosted at the University of Göttingen at `http://ecrit.net.informatik.uni-goettingen.de/wp/`.

A similar project by Dr. Krysztof Rzecki's group at Krakow Polytechnic University can be found on SourceForge at `http://ecrit.sourceforge.net/`. (To download the code, you'll need to go to the main SourceForge site, `http://sourceforge.net/projects/ecrit/`.) The Krakow server is written in Java, although no source code seems to be available as of this writing, while the client is written in C.

6.1.7 Zap! with Emergency Calling Extensions

Zap! is a universal SIP client built on the MozillaXULRunner framework. By using this framework, Zap! is platform-independent and very easy to extend. (It is licensed under the Mozilla License, available at `http://www.mozilla.org/MPL/MPL-1.1.html`.) Software downloads (binary and source) and more information can be found at the Zap! website at `http://www.croczilla.com/zap`.

The Zap! SIP client was the starting point for the nic.at project focused on implementing an emergency calling client. The extended client created by this project is available under Downloads at `http://www.voip-sos.net/`. With the extensions added by nic.at, Zap! can now support positioning, location configuration, location-to-service mapping, emergency call detection, and SIP location conveyance:

- Positioning:
  - GPS.
- Location Configuration:
  - HELD.
  - DHCP.
  - LLDP-MED.
- Determining the responsible PSAP:
  - LoST.
- Recognizing an emergency call:
  - Dial string/emergency number recognition (using LoST).

- Location Conveyance:
  - SIP Location Conveyance (by value).
  - SIP Location Conveyance (by reference).
- Routing:
  - Direct to the PSAP.
  - Via a VoIP provider's SIP proxy.
- Displays location information from an incoming call (when used to answer an emergency call).

The Zap! client can thus be used to look at emergency calling from the perspective of the caller. A knowledgeable user can see the status of emergency calling functions (location, LoST mappings, etc.) in the "Emergency Services Status" window, shown in Figure 6.4.

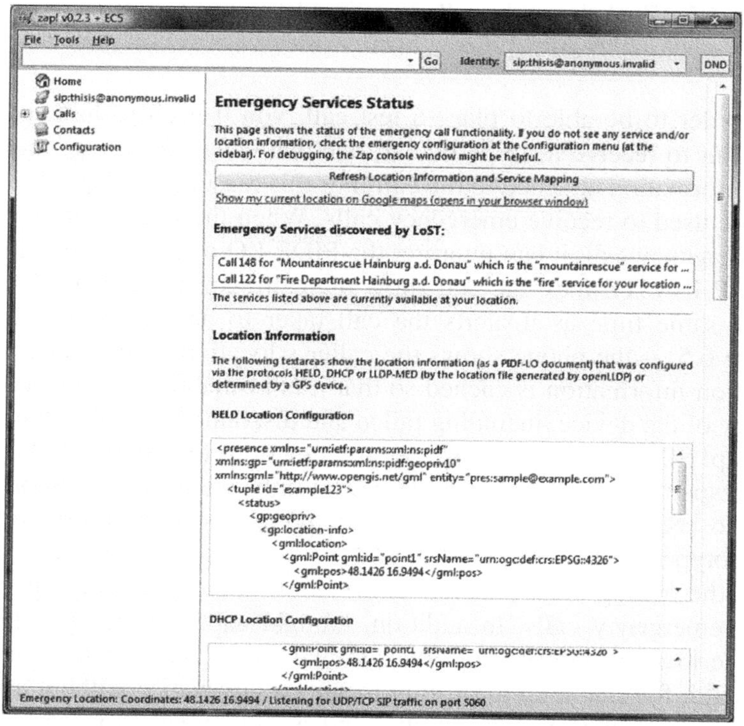

Figure 6.4 Emergency services status display in the Zap!-based emergency calling client. Zap! © 2005–2009 Alex Fritze – `http://www.croczilla.com/zap`.

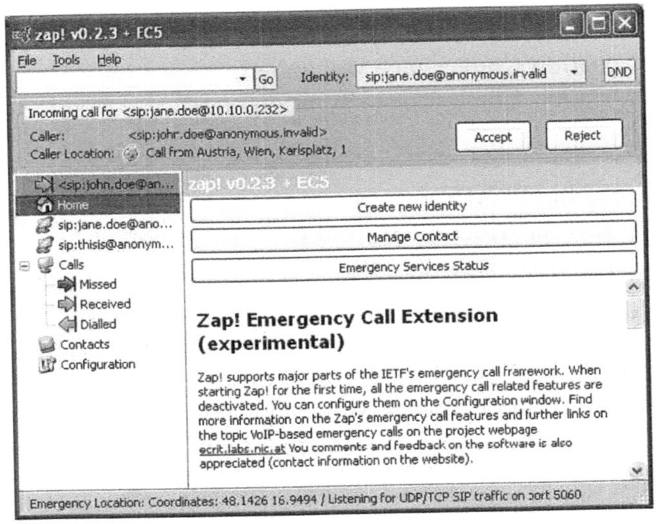

Figure 6.5 Zap! shows the caller location data carried in an incoming call. Zap! © 2005–2009 Alex Fritze – `http://www.croczilla.com/zap`.

In order to be able to place a test call, you'll need to have another endpoint to receive it, in particular, one that can make use of the location information attached to the call. For this reason, the Zap! client can also be used to receive emergency calls. When the Zap! client receives an emergency call, it can interpret the PIDF-LO document contained in the SIP INVITE message and show the call taker the caller's location at the same time as it alerts the call taker to an incoming call (see Figure 6.5) – the phone shows the caller's location as soon as it rings. Location information is cached so that it is available for the whole call history of the device, including failed and disconnected calls. Of course, the Zap! SIP client doesn't provide any functions to help the PSAP actually respond to emergencies, since it is only intended to demonstrate how to receive (emergency) calls including location information as a prototype.

On the calling side, location information is automatically appended to all emergency calls. In addition, though, callers have an option to include location information with normal calls. This functionality can be useful, for example, for testing location conveyance without a LoST server: The test user can simply send a call directly to a manually-configured PSAP.

When Zap! starts up, the client will try to acquire location information and information about responsible PSAPs. Some configuration information has to be set manually (see the Zap! documentation for details). Location information acquired using DHCP, LLDP-MED, or GPS will be converted into PIDF-LO format (location information from HELD is already in this format). If location information is available and a LoST server is configured, the LoST client will then try to acquire LoST mapping information. In order to first determine which emergency services are available at the caller's current location, it will first send a "listServicesByLocation" query. For each service listed in the response, the client will issue a separate "findService" query (as well as for urn:service:sos). Based on the mappings returned by these queries, it will create a cache of emergency dial strings and PSAP contact information. Once the client has this information, it can recognize when the user has dialed an emergency number.

Depending on which "identity" is selected in Zap! (i.e., with a SIP user account), the client either will place the emergency call using a SIP proxy provided by that identity's VoIP provider, or it will send the call directly to the PSAP. The user can refresh location and mapping information after startup (e.g., before an emergency call) by pressing the button labeled "Refresh Location Information and Service Mapping", shown in Figure 6.4.

The caller's location, as well as location information stored in the call history, can be displayed in a browser window by clicking on the Maps link that the client provides (the current code uses Google Maps, but of course, other web mapping services could be used with a simple modification to the client).

With the above functions, the Zap! SIP client makes it possible to place example emergency calls that use the ECRIT calling system, and to analyze the execution of an emergency call. Even in an environment with production-grade location, mapping, and calling services, the Zap! client can be a useful debugging tool, since it shows detailed information about the information these services provide.

6.1.8 Ecritdroid

To demonstrate emergency calling on a mobile device, Ecritdroid implements the ECRIT emergency calling architecture as an application for the Android smart phone operating system. Ecritdroid is open-source, released under the Apache license; the latest version can be downloaded from the project website at `http://ecritdroid.googlecode.com/`.

After Ecritdroid has been installed, it starts a background process that acquires location information from the operating system, then periodically refreshes LoST mappings based on this location information (it sends a LoST "findService" query for every Service URN). Based on these mappings, it creates a cache of mappings between emergency dial strings (the "serviceNumber" element in the LoST mapping) and PSAP contact URIs (the "uri" element in the LoST mapping). The user can view the available emergency numbers by tapping the application's icon (see Figure 6.6(a)).

Once the application has acquired a set of LoST mappings, the user can place emergency calls to IP-enabled PSAPs that have been discovered through LoST. Whenever the user dials a phone number (using the phone's normal dialer, Figure 6.6(b)), the operating system notifies the Ecritdroid application, which checks whether the number is an emergency number from LoST. If the number is an emergency dial string, then Ecritdroid presents the user with a list of contact URIs that correspond to the emergency number (Figure 6.6(c)). If the user taps any of the indicated URIs, then the operating system will open whatever application is registered to handle that URI type, thus initiating a call to the PSAP (Figure 6.6(d)).

Ecritdroid is interesting because it demonstrates that mobile networks are basically already ECRIT-enabled: As long as a handset has location information and an Internet connection, it can place IP calls with the right software. Modern smart phones have multiple ways to acquire location information, including GPS as well as network-based geolocation techniques such as those provided by the Google and Skyhook location services. And of course, almost all smart phones have Internet connectivity. So with appropriate software (such as Ecritdroid), basically all deployed smart phones can be upgraded to support ECRIT calling.

It's important to note, however, that Ecritdroid implements only the core part of the ECRIT architecture – using LoST to discover PSAPs. This leaves out two key components: Standards-based location configuration and location conveyance. Ecritdroid does not implement either DHCP or HELD, so it is unable to benefit from any specialized location services that a local network might offer. Instead, it has to rely on the general location services provided by the platform, whose performance can vary widely depending on the location of the device. The lack of location conveyance means that although the user will be able to contact a PSAP and speak with a call taker, the PSAP will not automatically receive the user's location information in the call signaling. Both of these issues could be resolved with some further software development, and may be addressed in later versions of Ecritdroid.

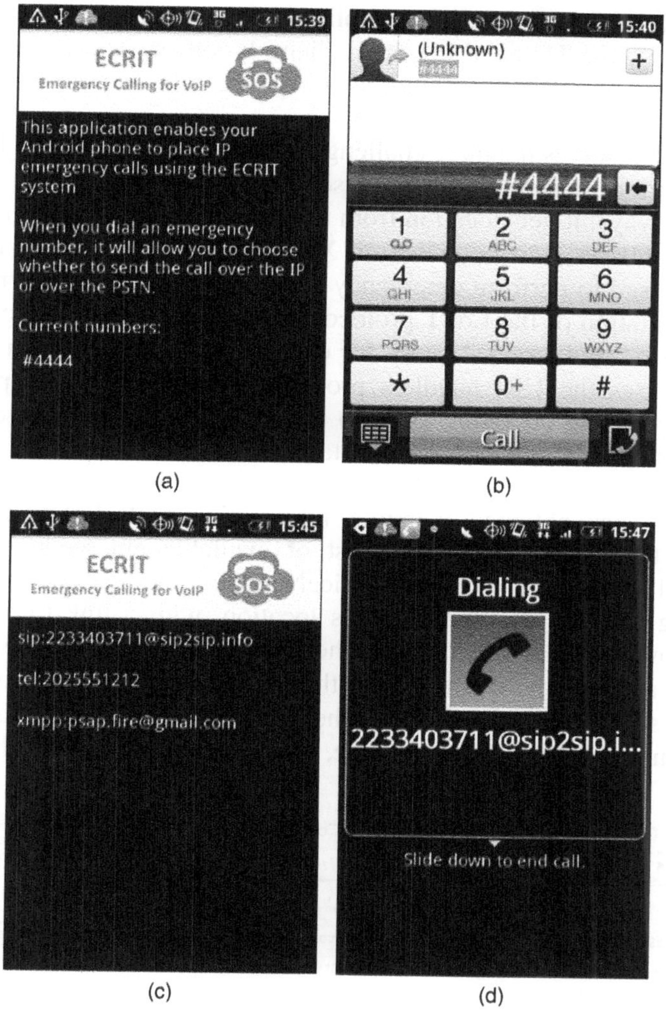

Figure 6.6 Emergency calling with Ecritdroid: Viewing available numbers (a), dialing an emergency number (b), choosing a PSAP URI to contact (c), and sending the emergency call itself (d). For more information, please visit: `http://ecrit.labs.nic.at`.

6.1.9 *EcritXUL*

Also part of the igtk suite of tools is the EcritXUL client, developed as part of the overall experimental set-up that we'll walk through in Section 6.2.6. EcritXUL is based on the XPCOM/XULRunner system that underlies several popular applications, such as the Firefox web browser and the Thunderbird email client. It comes packaged as a Firefox extension, but it can be run as a separate XULRunner application as well.

EcritXUL solves the three challenges of emergency calling – location configuration, emergency service discovery, and the call itself – in three different ways. It gets location from the igtk library, which has an interface that allows it to be called an XPCOM component. This location value is used to perform emergency service discovery using a minimal implementation of the LoST protocol in Javascript (given a location, it performs a "findService" query for each Service URN). When placing calls, it uses the URI handlers provided by the underlying platform (Firefox and the operating system) to invoke the appropriate communications applications for each contact URI.

The user interface for EcritXUL is shown in Figure 6.7. After EcritXUL has obtained mapping information with LoST, it builds a table that shows the user a list of available emergency services and dial strings (from the "serviceNumber" element in the LoST mapping). It also shows the user's location, with a link to a Google map showing the location. When the user clicks to expand the row for a given service, the client shows the contact URIs that are available for that service. In this example, the local fire department is available by calling 911 on a normal phone, or by sending a call to the given

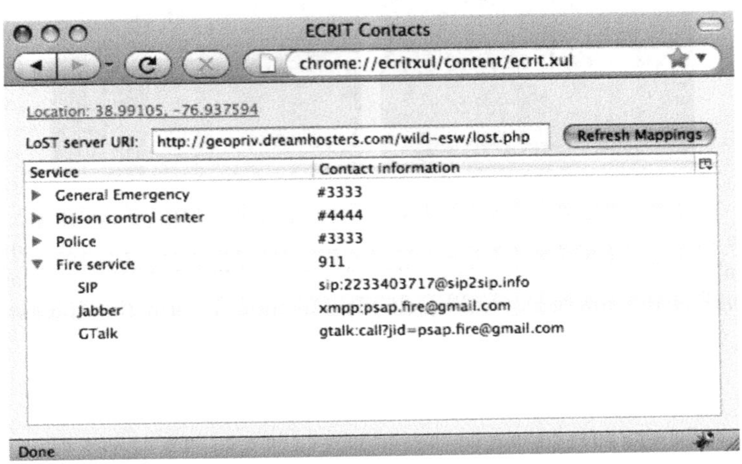

Figure 6.7 EcritXUL showing available emergency services, dial strings, and contact URIs. For more information, please visit: `http://ecrit.labs.nic.at`.

SIP, XMPP, or GTalk URIs ("gtalk" is a URI scheme used by the Google Talk client). When a user clicks on one of the URI rows, the appropriate application is opened to initiate communications with the PSAP via that URI. For example, if the user has a SIP client installed, clicking the SIP URI in the windows shown in Figure 6.7 would open that client and start a SIP call to the PSAP.

Like Ecritdroid, however, it should be noted that EcritXUL does not perform any special actions at the application layer. For example, in SIP, there will be no Service URN in the call signaling, and no location conveyed. So while this client will send a call to a PSAP, it will not mark the call in a way that the VoIP provider can recognize it as an emergency call, and it will not provide automatic location.

6.1.10 Multi-Part Body Extension to Asterisk

One important difference between normal calls and emergency calls in SIP is that emergency calls will often have a multi-part body in the SIP INVITE message that initiates the call, with one body part carrying the normal SDP information to set up the session and another carrying a PIDF-LO document with the caller's location. So SIP intermediaries (e.g., proxies and SBCs) will need to be able to process multi-part bodies.

In particular, many proxies will need to access location objects carried in SIP in order to do location-based call routing. For example, a proxy might extract the location object, perform a LoST query to determine the correct PSAP URI, then redirect the call to that URI. In order to support such cases in the popular Asterisk PBX software, nic.at has submitted a patch to Asterisk that extends the current call routing functionality so that the "dial plan", the part that determines routing based on a SIP INVITE message, can examine multiple body parts from a multi-part MIME body in an INVITE message. For instance, a dial plan that performs location-based routing would be able to make a call of the form:

```
${SIP_BODY(application/pidf+xml)}
```

This call would then return the PIDF-LO content of a message (which has content type "application/pidf+xml").

This patch has not yet been accepted into the main Asterisk code base, and it may change before being accepted. Further information on the status of the extension and the patch itself can be found on the Asterisk Issue Tracker and on this book's website, at `http://www.voip-sos.net/tools/asterisk/`.

Also, be aware of an Asterisk specific limitation: Asterisk limits the number of lines of text in a SIP message body that it will parse. Old Asterisk releases can only handle bodies with 64 lines or less. Hence, whenever an Asterisk receives a multi-part body that is longer than this limit, the call will fail. Large PIDF-LO documents can easily exceed this limit, especially in combination with SDP bodies. Recently, this limit was increased to 256 lines – an improvement, but still not a satisfactory solution (think about multiple PIDF-LO elements in the body). If you run into such troubles and Asterisk claims not to be able to find your desired body part, you can increase the line limit by modifying and recompiling Asterisk. Look at the Asterisk source file *channels/sip/include/sip.h* for the following line:

```
/*!< Max amount of lines in SIP attachment (like SDP) */
#define SIP_MAX_LINES 64
```

Increasing this number will allow Asterisk to process larger message bodies.

6.1.11 IMS Core Emergency Services

It is likely that many SIP emergency calls will be placed using networks configured according to the IP Multimedia Subsystem (IMS) specifications published by the 3GPP. As part of the EU PEACE program, the Fraunhofer FOKUS Institute in Germany has extended their Open IMS Core to implement the emergency calling system defined for the IMS.

The basic Open IMS Core provides implementations of several critical IMS functions, such as the Home Subscriber Server (HSS) for user tracking and SIP proxies known as Call Session Control Functions (CSCFs). The emergency branch of the Open IMS Core extends this basic core network by adding two emergency-related entities, an Emergency CSCF (E-CSCF) and a Location Retrieval Function. The E-CSCF is simply a SIP proxy that routes emergency calls to the proper PSAP, using information obtained from the LRF.

The LRF provides location-based call routing information: It receives information about the caller's location either in the request from the E-CSCF or from a separate location service. This caller location is

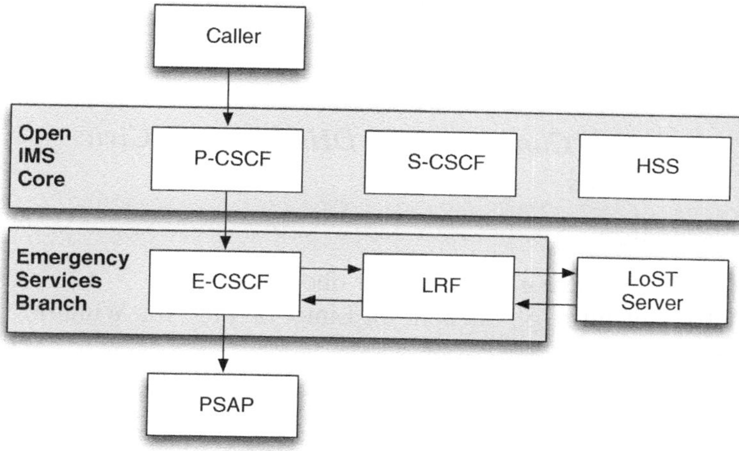

Figure 6.8 Elements of the emergency branch of the FOKUS Open IMS Core.

included in a LoST "findService" query (using a Service URN provided the E-CSCF), and the resulting PSAP contact URI is provided to the E-CSCF. Once the E-CSCF receives the PSAP URI from the LRF, it forwards the call to the PSAP.

The emergency services implementation in the Open IMS Core can be used to represent the core signaling network in VoIP emergency calling experiments. For example, you could set up the Open IMS Core in a network with a HELD server and a LoST server (using some of the open-source servers mentioned above), then use the Zap! emergency calling client for the client and PSAP terminals.

Figure 6.8 illustrates these components and how they interact. More information about this implementation of IMS emergency services can be found on the Open IMS Core website at `http://openimscore.org/emergency`.

6.2 Practice Exercises

In the following sections are a few experiments you can use to explore VoIP emergency calling, using the software described above. No software needs to be purchased for this test environment, since it is based entirely on open-source software. The steps here are deliberately simple to allow the reader to gradually become familiar with the key themes. For steps involving the Zap! client, the user manual for the Zap!

emergency calling client can also be helpful (available at `http://www.voip-sos.net`).

6.2.1 *Location Configuration: DHCPv4 with Civic Addresses*

- **Prerequisites:**
  - DHCP server (e.g., dnsmasq or dhcpd).
  - DHCP client (e.g., dhclient on Linux or Zap! for Windows).
  - Wireshark.
- **Goal:** Familiarity with DHCP location configuration.

The section describes how the DHCP location options can be used for location configuration in a small network. For this function, we'll use a civic address, carried in DHCP option 99 (as defined in RFC 4776).

The underlying assumptions for this case are that the network in question uses a DHCP server for address management, and that it serves a specific geographical area, so that all devices in the network can be configured with the same address. This geographical area might be as small as a single building (or even a single room), or as large as a neighborhood or city, as long as the area is sufficiently small that it is only served by one set of emergency responders. The overall point here is that location in support of emergency calling doesn't necessarily have to pinpoint the caller, only locate him precisely enough to get to the right PSAP; this flexibility can make it easier to deploy location services that are good enough for placing emergency calls.

Note, however, that even though coarse-grained location is good enough for placing emergency calls, it may not be good enough to direct responders to the scene of an emergency (e.g., a floor or room within a building). So while this section describes a basic way of providing clients with a location that is good enough to get a call through, real deployments should try to provide more precise location that can be used to dispatch responders to the emergency more precisely.

The first step is to take a location value and encode it in the option 99 (RFC 4776) format. The simplest way to do this is to use one of the available online encoders (see Section 6.1.2 as well as the companion website at `http://www.voip-sos.net`). However, in order to use these tools, you will have to figure out which elements of the address correspond to the different civic address elements. Descriptions of the elements are found in RFC 4776 and RFC 5139, and additional nation-specific considerations may apply (see Section 3.2.1). Since we're configuring endpoints with location information that is broad enough to cover the endpoint as well as network elements and the DHCP server, any value of the "what" field is appropriate.

Once the location information is encoded in the proper format, it will need to be provisioned into the DHCP server. Since option 99 is not supported as a native option on current DHCP servers, you will need to use your DHCP server's commands to define a new option and set the value of that option. For instance, if you are using the ISC DHCP server (dhcpd), you might use the following configuration lines:

```
option loc-civic code 99 = string;
option loc-civic 00:41:54:01:04:57:69:65:6e;
```

Likewise, a line of the following form can be used with the dnsmasq server on Linux:

```
dhcp-option=99,01:41:54:01:04:54:65:73:74
```

After these lines have been added to the DHCP server's configuration files, the server will need to be loaded on the configuration. From that point, the DHCP civic location option will be provided to clients. For existing clients that don't understand option 99, there will obviously be no change; in order to expose the DHCP location information to applications, you will need to apply some platform-specific changes, described for Windows and Linux below. On the other hand, you can view the DHCP location information directly by using the Wireshark protocol analyzer. After you have captured some packets with Wireshark, you can have it display only the DHCP packets by setting the "bootp" display filter. Since Wireshark has a decoder for option 99, your location information will show up in the packet dissection window, as shown in Figure 6.3.

6.2.1.1 Requesting DHCP Options with dhclient3 under Linux

In order to use the standard Linux DHCP client "dhclient3" to query the DHCP server for option 99, you just need to make a small modification to the client configuration file (e.g., `/etc/dhcp3/dhclient.conf`).

Namely, you should add the value “unknown-99” to the “request” line in the configuration field, for example:

```
request subnet-mask, broadcast-address, time-offset,
routers, domain-name, domain-name-servers, host-name,
netbios-name-servers, netbios-scope, interface-mtu, nwip-
suboptions, unknown-99;
```

When the DHCP client next starts up, option 99 will be requested from the DHCP server.

6.2.1.2 Requesting DHCP Options with Zap! under Windows

In order to view the location information in DHCP options in a setting other than Wireshark – and also to make more productive use of it – you can also use the Zap! SIP client with the emergency calling extensions. The extended version of Zap! client can be downloaded at `http://www.voip-sos.net/tools/zap/`. After downloading software and expanding the archive, start the Zap! client by opening the application “zap.exe”. The first time that Zap! starts up, DHCP functions will be deactivated. In order to activate them, open the “Configuration” window and activate the option “Discover DHCP location” under “Emergency Calls”, then click “Apply Changes”. Once this configuration change has been made, DHCP location configuration can be activated by restarting Zap!, or just by clicking on the button labeled “Refresh Location Information and Service Mapping” in the “Emergency Services Status” window. The configured location should appear at the left bottom line of Zap!. The PIDF-LO that Zap! has constructed based on the DHCP location information (which will be sent to the PSAP in an emergency call) can be seen in the “Emergency Services Status” window. Next to it, there is a Google Maps link that will show the configured location on a map. Of course, you can also verify that DHCP information has been delivered to Zap! by watching for it in Wireshark as well.

After these steps, Zap! has been configured with location information via DHCP and can then make further use of this information; for example, it can use the location information to request LoST mappings.

6.2.2 *Location Configuration: Simulating a HELD Server*

- **Prerequisites:**
  - Web server (e.g., apache).
  - HELD client (e.g., Zap!).
- **Goal:** Simple location configuration with HELD.

The HELD protocol has several functions, and there is already some work underway to extend it. In order to quickly create an experimental set-up that supports location configuration with HELD, it's not necessary to create a complete Location Server. The basic function of the protocol – providing location information, without accepting parameters from the client – requires only an HTTP POST request, which is answered with a HELD location response XML object (Barnes et al., 2010). (In principle, the POST should contain a HELD request XML body, but it isn't strictly necessary for the protocol to work. There is actually a document proposed to allow HELD to work with GET requests, without an XML body to the request (Thomson, 2010).) So for test purposes, you can create a basic HELD server simply by provisioning such an XML object on a web server. Such a Location Server would of course have no positioning capability, since the same location information would be delivered to all clients. Nonetheless, using a basic web server to provide a HELD location responses is a quick and easy way to allow location configuration via HELD.

In more concrete terms, you will need to construct a HELD "locationResponse" XML document and store it on your web server. The content of a location response is described in Section 4.1.1, and example responses are available at `http://www.voip-sos.net/tools/pidflo/`, and in the next section of this book.

Of course, only HELD clients can use this Location Server for location configuration. One such client is included in the emergency extensions to the Zap! client, available at `http://www.voip-sos.net/tools/zap/`. After downloading and starting the Zap! client, you will need to open the "Configuration" window in Zap! and set the "Location Server" setting to the URI for the Location Server, in this case, the URI where you provisioned the HELD location response. Clicking "Apply changes" will store the required configuration changes. Zap! will request location information from the Location Server when it is restarted, or when the "Refresh Location Information and Service Mapping" in the "Emergency Services Status" window is clicked. The location information will be displayed in the bottom-left portion of the Zap! window, and the full PIDF-LO document will be shown in the "Emergency Services Status" window. If location configuration fails, you can look at the HELD request in Wireshark by setting the display filter to "http && ip.addr == [ip-addr]" where "[ip-addr]" is the IP address of the Location Server.

As described in Section 6.1.1, there are also several open-source HELD servers that are not much harder to set up than a static web page. So as an advanced exercise, you could download and install one of those HELD servers, or try out one of the HELD clients discussed above. Once the server is installed, the process for querying the server and looking at the results in Wireshark is exactly the same as above.

6.2.3 Location Configuration: Location-Enabling a Network with HELD

- **Prerequisites:**
  - Web server (e.g., apache).
  - MySQL database.
  - HELD client (e.g., Zap!).
  - DHCP and DNS servers (optional).
- **Goal:** Dynamic IP-based location configuration with HELD.

In this exercise, we will go through all the steps needed to location-enable an IP network: We will set up a basic HELD Location Information Server (LIS), supply it with location information, and supply the DNS and DHCP records necessary for clients to discover it. The LIS will return location information from two sources, based on the IP address of the client: If it has been supplied with location information for the client's IP address, then it will return that location information; otherwise it will return location from the MaxMind geo-IP database. Location will be supplied by associating location information with an IP address prefix, so that any IP address that begins with that prefix will be provided with the supplied location information.

This setup could be used directly in a network where different subnets are associated with different locations. For example, consider an enterprise with three locations, each of which has a subnet of the corporate network assigned a different IP address prefix. This enterprise could easily create a Location Server for their network by following the instructions in this section to install the supplied LIS, then supplying the LIS so that the prefix for branch office is associated with the address of that branch office. Once the Location Server was set up, the enterprise would only need to install the DHCP and DNS records for LIS discovery in order to be fully location-enabled.

The first step in this exercise is to download the LIS software package, which is available at `http://www.voip-sos.net/tools/held/`. This package is based on the Perl LIS from the igtk project, and comes with all the necessary scripts to install, run, and provision the LIS. While that downloads, collect the following information:

- Log-in details for your MySQL database, and the name of the database to use.
- A Google Maps API key (you can get one for free if you don't have one).

In the LIS software package, the README file will guide you through the precise steps to set up the LIS. You will need to run a SQL script to set up the appropriate tables and edit the scripts for the LIS and the provisioning page to insert the proper database details and the API key. Once the database and the scripts are set up, all you need to do is upload the scripts to your web server (the LIS script will need to be in a place where it can run as a CGI script). For the rest of this exercise, when we refer to the "LIS URI" it will be the location where you have installed the LIS script; for illustrative purposes, we'll assume the LIS URI is `http://example.com/lis/lis.pl`.

Once you've installed the LIS and provisioning scripts, you should immediately be able to test the LIS. You can use any of the HELD clients described above to access the LIS, but for convenience, the LIS package also comes with a built-in HELD client that runs as a web page. To use that client, install the "viewer" script from the LIS package (instructions are in the README), and simply access the script on your web server using a standard web browser. Once you fill in the URI for the LIS and press "HELD Request", the viewer should perform the HELD query and show the location returned for your IP address by the LIS. Since we haven't supplied any location information yet, this location will come from MaxMind, and might be inaccurate. But once we supply it, the LIS will return the supplied information, and once we install LIS-discovery records, the viewer should be able to automatically find the LIS. (Note, however, that since the viewer uses an XMLHttpRequest for HELD, the viewer will only work if it's on the same server as the LIS!) Of course, whichever HELD client you use, you can always use Wireshark to examine the HELD requests on the wire.

Now that we can pull data from the LIS, let's supply it with some location information. For this, you'll need two things: A prefix and a location value. If you happen to know which prefix your local network uses, you can use that one; otherwise, just use the client computer's IP address as a prefix of length 32. For the location value, you obviously choose any value you like.

To provision the LIS, simply open the provisioning page in your browser and enter the information. The top of the provisioning page should look as shown in Figure 6.1 above. The first thing to enter is the prefix, which we enter in "prefix/length" form (e.g., `192.0.2.0/24`), with separate fields for the prefix and the length. To enter geodetic information, use the map tool to specify a point, circle, rectangle or polygon (or click "Clear" to start over). To enter civic information, simply fill in the forms to specify values for the different parts of the civic address.

Once you've entered a prefix and a location value, click "Add location" to implement the location into the LIS database. The location should immediately be shown in the list of supplied locations at the bottom of the provisioning page. Now, when you access the LIS from any of the IP addresses for which you've supplied location (using any HELD client), you should get the supplied location from the database.

After all of the above steps, you now have a HELD LIS that clients can use to discover where they are. The next step is to enable clients to find the LIS when they connect to the network. This step will require that you be able to implement records into the DHCP server for the network, and into a DNS server (not necessarily bound to the network). If you don't have access to either of these, you can either create a small test network (as described in Section 6.2.6), or you can just skip this step and manually configure HELD clients.

Recall that the LIS discovery process has two parts: First, the client gets a domain name from DHCP, then it retrieves NAPTR records for that domain name in order to find the LIS URI. Likewise, there are two steps to enabling LIS discovery, supplying a DHCP record and supplying DNS records. Recall that for this example, we're assuming that the LIS URI is `http://example.com/lis/lis.pl`, and the domain name for the local network is `example.com`.

For DHCP, you need to add a line to your DHCP server configuration that tells the server to send the LIS discovery option to all clients. Most DHCP servers will require that this option be supplied in hex form. The encoding for this option is the same as for other domain name options (e.g., option 15); labels are sent in ASCII, with each label preceded by a byte with the length of the label, plus a null terminator. So if you wanted to encode "`example.com`", the bytes would be as follows:

```
7 e x a m p l e 3 c o m 0
```

Encoding the ASCII bytes in hex, we get the full hex string

```
07:65:78:61:6d:70:6c:65:03:63:6f:6d:00.
```

So, for example, in order to implement this option into a Cisco DHCP server, you would add the following line to your DHCP pool configuration:

```
option 213 hex 07:65:78:61:6d:70:6c:65:03:63:6f:6d:00
```

Implementing the DNS is simpler; you just need to edit your zone file to add a few records. The first record to add is for the name we just implement into the DHCP option. Using the BIND configuration format, the NAPTR record has the following form:

```
example.com  IN NAPTR 100 10 "u" "LIS:HELD" \
             "!.*!http://example.com/lis/lis.pl!"  "".
```

Obviously, you'll be using a different domain and URI, so just replace those portions of the record, that is, the name at the beginning and the URI between the exclamation points.

It's also a good idea to implement these records in the reverse DNS for the network in question, but there's currently not any support for NAPTR records in standard tools for generating reverse zones. If you want to create these reverse records, you'll need to explicitly add the record for each address, but this is pretty easy to script. Just output one line for each address, with the reverse name for the address followed by the content of the NAPTR record (which is the same for every record).

This approach can get to be cumbersome for larger networks, but works reasonably well for smaller networks (around a /24).

With your LIS discovery records in place and your LIS set up and implemented, hosts on your network can now use standard HELD clients – such as the igtk client, the Andrew open-source client, or even Firefox – to get access to their location information.

6.2.4 *Mapping: Querying the LoST Server*

- **Prerequisites:**
  - Web server (e.g., apache).
  - Zap! emergency calling client.
  - Wireshark.
  - LoST server (optional).
- **Goal:** Simple location-to-service mapping with LoST.

Finding the appropriate PSAP for the caller's current location is a critical challenge for emergency calling. In order get basic familiarity with the mapping process enabled by the LoST protocol, you can use the LoST web client from Columbia University (see Section 6.1.6). The web client will allow you to send queries to a sample LoST server and see the responses. In this section, we describe another approach, using the Zap! client with extensions for emergency calling.

First, you will need to download Zap! from `http://www.voip-sos.net`, unzip it, and start it up. In order to perform mapping queries, Zap! will need access to location information. To provide this information, store the following HELD "locationResponse" object (with the coordinates (35.2225, -80.8449), a location in Charlotte, North Carolina) on a web server that the Zap! client can access:

```
<?xml version="1.0" encoding="UTF-8"?>
<locationResponse
xmlns="urn:ietf:params:xml:ns:geopriv:held">
<presence xmlns="urn:ietf:params:xml:ns:pidf"
   xmlns:gp="urn:ietf:params:xml:ns:pidf:geopriv10"
   xmlns:gml=
     "urn:opengis:specification:gml:schema-xsd:feature:v3.0"
   entity="pres:sample@example.com">
  <tuple id="xyz">
   <status>
```

```
      <gp:geopriv>
        <gp:location-info>
          <gml:location>
            <gml:Point gml:id="point1"
               srsName="urn:ogc:def:crs:EPSG::4326">
              <gml:pos>35.2225 -80.8449</gml:pos>
            </gml:Point>
          </gml:location>
        </gp:location-info>
        <gp:usage-rules/>
        <method>wiremap</method>
      </gp:geopriv>
    </status>
    <timestamp>2011-01-01T04:41:59Z</timestamp>
  </tuple>
 </presence>
</locationResponse>
```


In the Zap! “Configuration” window, fill in the URI for a Location Server, for example, the URI where you stored the “locationResponse” document in the last step.

In addition, in order to do mapping, you will need to specify a LoST server. You can install and provision your own using the Columbia University LoST software (see Section 6.1.6), or you can simply use the Columbia demonstration server at `http://ng911serv.irt.cs.columbia.edu:8080/lost/LoSTServlet`. The location in the location response above is one for which the Columbia server should return a mapping. If you create your own LoST server, you might need to change the location value in the location response to one that your LoST server knows about.

After configuring a Location Server and a LoST server, you need to click the “Apply changes” button at the bottom-right of the Zap! window. Once the changes have been applied, the Zap! client will be able to acquire location information and LoST mappings. After these two pieces of information have been collected, the “Emergency Services Status” window will display the PIDF-LO document with the location information as well as information about the emergency services that it has discovered through LoST. The Zap! window should look like the window shown in Figure 6.4. If no emergency services are discovered, check that location information was configured and that the location value is one that the LoST server knows about (e.g., the location value shown above).

Communications with the HELD server and the LoST server can both be analyzed using Wireshark, since they both use XML documents transported in HTTP. To view just HELD or LoST transactions, set the display filter to "http && ip.addr == [ip-addr]" where "[ip-addr]" is the IP address of the Location Server or the LoST server (to view both, use "http && (ip.addr == [held-ip-addr] || ip.addr == [lost-ip-addr])"). To view the contents of an HTTP transaction in a consolidated view, select one of the HTTP messages in the transaction and select the "Follow TCP stream" option under the "Analyze" menu.

In addition, you can also start up Zap! with the parameter "-console". This option will cause Zap! to open an additional window that shows communications with the HELD server and the LoST server. This extra information can be very useful for debugging.

Of course, this example setup can also use other location values or other techniques for location configuration. (Running this example with location information provided over DHCP is left as an exercise to the reader.) As noted above, you can also install your own LoST server, which has one particular advantage: The SIP URIs that the Columbia LoST server provides as contact information for PSAPs are not actually functional; they don't respond to calls. If you run your own LoST server, you can test the entire emergency calling process, using URIs from LoST to place real SIP calls.

One final note: In this scenario, we only discussed emergency services, but the LoST server can of course provide mappings for other services. They can be supported by clients similar to emergency calls, in particular the same protocols can be used.

6.2.5 SIP Calling: Call Setup with Location Configuration

- **Prerequisites:**
  - 2 Zap! clients.
  - A location configuration service (DHCP or HELD).
  - Wireshark.
- **Goal:** Familiarity with location configuration.

With the extended Zap! SIP client, it is also possible to convey location information in the SIP INVITE messages that signal the beginning of a VoIP call. For emergency calls, the Zap! client appends this information automatically, and for normal calls, the caller has the option to append location information. This feature can be useful for testing emergency calling without LoST, since if you already know the URI for a PSAP, then you can simply use Zap! to acquire location information and include it in a call to the PSAP.

Of course, in order for Zap! to convey location in an INVITE message, it needs to have a location in the first place. So you will need to deploy one of the location configuration services described above (DHCP or HELD), or use a device that provides GPS information to Zap!.

In order to receive a call with location information, you will need a second Zap! client (again with the emergency calling extensions) that emulates a PSAP. This second installation doesn't need access to the location configuration service, since it will get location from the caller. But the Zap! emergency calling extensions are needed to interpret the location included in the call.

In order to place a call with Zap!, you only need to enter the SIP address of the destination in the address bar. In order to place a call to a client on your local network, you just need to enter the local IP address of the destination in the address bar. If you want to perform call setup through a SIP proxy, you will need to configure an "Identity" in Zap!, but otherwise you can place the call directly without configuring an identity (this will be indicated by the anonymous identity "anonymous.invalid" next to the address bar).

After entering a SIP destination and clicking "Go", a call setup window will appear, as shown in Figure 6.9. Since this is not an emergency call, location information is not conveyed automatically. However, you can tell Zap! to add location information by selecting the box marked "I want to convey location information for this non-emergency call to the remote party!". If Zap! detects an emergency call (based on the dial string), then it will automatically add location information. You can try this when you have valid LoST mappings – the dial strings are then shown on the "Emergency Services Status" window (as described in Section 6.2.4) and have only to be entered instead of an SIP URI in the address bar. After clicking the "Call" button, the SIP INVITE message will be sent and the call will be established.

On the receiving end, the other party's Zap! client should now present the incoming call along with the location information it contains (as shown in Figure 6.5). If the recipient accepts the call, the location information field will be presented in the "Location" field in the "Active

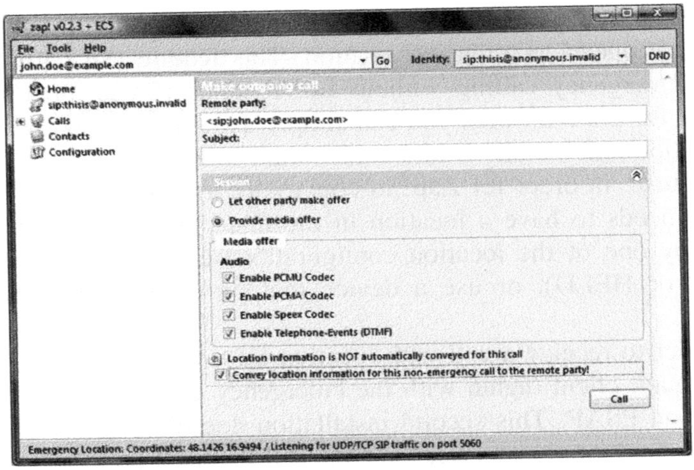

Figure 6.9 Location information can optionally be added to outgoing calls. Location information is automatically attached to emergency calls. Zap! © 2005–2009 Alex Fritze – `http://www.croczilla.com/zap`.

Inbound Call" window (see Figure 6.10). Clicking on the word "Location" will open a browser window that shows the location on a map. The location displayed to the recipient of the call (whether in Zap! or on the map), should be the exactly the same as the location provided to the client by the location configuration service (e.g., through DHCP or HELD).

After the call is ended, the location information can still be viewed in the call window. Location information is also stored in the call history, from which it can be recalled later on. To access location information for a call, click the "SIP Logs" button in Zap! (see Figure 6.11). In the call log window that opens, you can review the SIP messages that were sent during the call. Location information is indicated by the "Geolocation" header, and by a PIDF-LO body in the SIP message (referenced by the "Geolocation" header), as described in Section 3.2.3. If there is a problem with call setup, the SIP messages sent by Zap! can be analyzed with Wireshark. To isolate SIP messages in a Wireshark view, set the display filter to "sip". (You can also use the "ip.addr" filter to look for SIP traffic to specific IP addresses.) In addition, it is frequently helpful to first try placing a call without sending location information. Finally, since Zap! lacks a robust mechanism for working through NATs, this example works best if both Zap! installations (caller and callee) are on the same network.

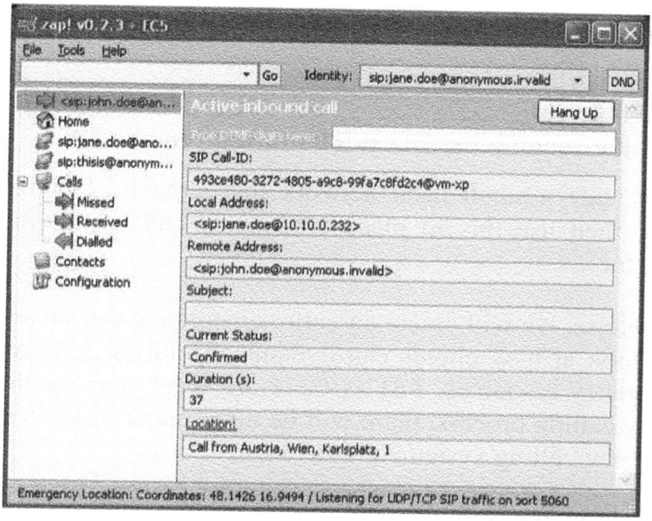

Figure 6.10 Location information is shown for an active inbound call. Clicking on "Location" opens a browser window with a Google Map for the indicated location. Zap! © 2005–2009 Alex Fritze – `http://www.croczilla.com/zap`.

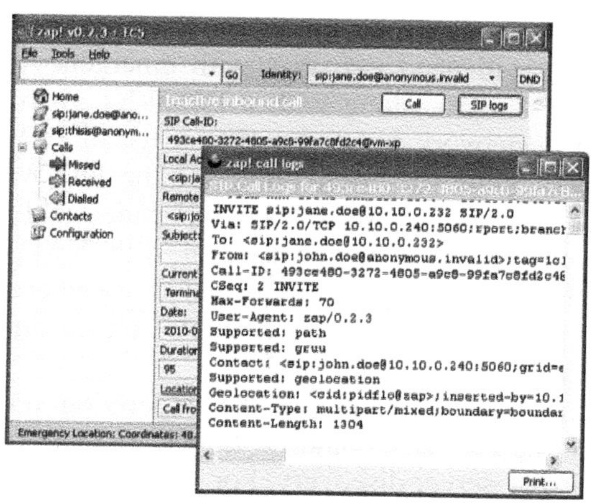

Figure 6.11 After the call, you can examine the SIP logs. Zap! © 2005–2009 Alex Fritze – `http://www.croczilla.com/zap`.

6.2.6 *ECRIT Calling: A Complete System*

- **Prerequisites:**
  - 2 calling clients (e.g., Zap!).
  - A web server.
  - A router running the DD-WRT firmware (optional).
- **Goal:** Create a small local network that supports location configuration, then use this to place emergency calls using the ECRIT architecture.

In this exercise, we will create an entire emergency calling system and use it to look at how clients can dynamically discover local location and emergency resources. Just as there are three steps in the ECRIT emergency calling process, there will be three major tasks in setting up an emergency calling system:

- Location-enabling the network.
- Creating a LoST server.
- Setting up a PSAP to receive calls.

We'll go about these out of order. First we'll set up the network and the VoIP soft phones, then we'll create the LIS and the LoST server. Once these configuration steps are complete, we'll be able to place emergency calls.

The overall network architecture or this project is shown in Figure 6.12. You will need two PCs with emergency calling clients (such as the Zap! client), along with a DD-WRT enabled router that will provide local network functions. (For a list of DD-WRT capable routers and instructions for installation, see the DD-WRT website at `http://www.dd-wrt.com/`). We'll keep the caller and the PSAP on the same network to avoid having to deal with NAT issues in SIP. The HELD and LoST servers can be installed anywhere, either on the LAN served by the router or somewhere else on the Internet, as long as they're accessible by the caller PC.

The first thing to do is configure the underlying IP network, and in particular, set up the DHCP and DNS systems with records for LIS discovery and LoST server discovery. For the purposes of this exercise, we'll assume that the network is configured as follows:

- LAN Domain: access-network.local.
- LIS URI: `http://example.com/lis/`.

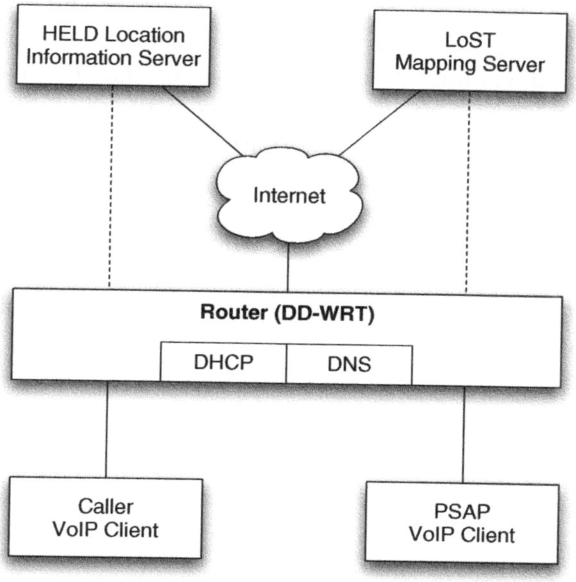

Figure 6.12 Network layout for a complete ECRIT experiment.

- LoST URI: `http://example.com/lost/`.
- Static DHCP lease to caller PC:
  - MAC Address: (MAC address of caller PC).
  - Host Name: caller.
  - IP Address: 192.168.1.100.
- Static DHCP lease to PSAP PC:
  - MAC Address: (MAC address of PSAP PC).
  - Host Name: psap.
  - IP Address: 192.168.1.911.

(Note that you may need to change the addresses in these leases to fit within the subnet that the DHCP server is configured to use.)

To configure the router with this information, you can use the web-based configuration interface for DD-WRT. Open the configuration page on the router with a browser and log in. Under the "Basic Setup" section of the "Setup" tab, most of the settings should be OK by default, but make sure that "Use DNSMasq for DHCP" and "Use DNSMasq for DNS" are both checked. In the "DHCP Server" section of the Services tab, we'll enable the DHCP server and configure it to send the right option values. Set the "Used Domain" value to "LAN" to enable local configuration of the domain that the DHCP server uses. Set the LAN domain to the access network domain name of your choosing (e.g., access-network.local, as above), and enter the information for the DHCP leases.

In the "DNSMasq" section, we'll enable the DNS server and add the DHCP options and DNS records. Obviously, you'll want to have the "DNSMasq" and "Local DNS" options enabled. The configurations for the DHCP and DNS records go into the "Additional DNSMasq Options" box:

```
local=/local/
expand-hosts
dhcp-option=213,access-network.local
naptr-record=access-network.local.,10,100,u,LIS:HELD,\
!.*!http://example.com/lis/!
dhcp-option=137,access-network.local
naptr-record=access-network.local.,10,100,u,LoST:http,\
!.*!http://example.com/lost/!
```

Let's go through these configuration lines one-by-one. The first two lines configure the local DNS to use the local domain from DHCP as the local domain for DNS (i.e., the "access-network.local" domain that we set up above), and append that domain to host names, so that we can use, for example, `psap.access-network.local`. The third through fifth lines set up discovery records for HELD, first advertising the domain name in DHCP then adding the corresponding NAPTR record with the LIS URI. Lines six through eight do the same thing for the LoST server.

Now that we have the network running, we'll install and provision the LIS and the LoST server. To install the LIS, follow the instructions in Section 6.2.3; once it's installed, supply it with a location for this network. For the LoST server, you can use any of the open-source LoST server implementations discussed in Section 6.1.6 above. The LoST server should be supplied with a mapping of the following form:

- Service boundary: Any region containing the location provisioned in the LIS. The easiest thing to do is probably to include the whole world.
- Service URN: urn:service:sos.

- Service number: 911.
- URI: `sip:psap@psap.access-network.local`.

If you want to avoid the trouble of installing a full LoST server, you could also simply make a static LoST "findServiceResponse" document available on the LoST server, such as the following:

```
<?xml version="1.0" encoding="UTF-8"?>
<findServiceResponse
  xmlns="urn:ietf:params:xml:ns:lost1"
  <mapping expires="2012-01-01T01:44:33Z"
    <service>urn:service:sos</service>
    <uri>sip:psap@psap.access-network.local</uri>
    <serviceNumber>911</serviceNumber>
  </mapping>
</findServiceResponse>
```

At this point, you should be location-enabled and LoST-enabled, and you should be able to place VoIP calls directly between the caller and the PSAP. You should be able to get location information from the LIS and mapping information from the LoST server using standard clients, as discussed in Section 6.2.3 and Section 6.2.4 above.

To establish VoIP connectivity, all we really need to do is tell each phone who it is, and make sure the phones can call each other. On the PSAP PC, open the Zap! client and click "Create new identity". In the "Address of Record" field, enter "sip:psap@psap.access-network.local", and un-check the "Automatic Registration" box. On the caller PC, do the same thing, but set the address of record to "sip:caller@caller.access-network.local". To test that the two phones can talk to each other, go the caller's Zap! client and place a call directly to the PSAP's URI.

The final step is to perform the actual emergency call! On the caller's PC, you'll need to configure the Zap! client with the URIs for the LIS and the LoST server. In the "Emergency Calls" portion of the configuration screen (click "Configuration" in the panel on the left), enter these URIs into the "LIS URI" and "LoST Server URI" fields, respectively. Also, since we haven't configured DHCP location, make sure this is unchecked.

(This configuration step is necessary because the Zap! client unfortunately doesn't yet support LIS discovery or LoST discovery (as of this writing). Some clients, such as the igtk client and EcritXUL, already support LIS discovery, however, so you can test that functionality in this network.)

Once you've set up the emergency configuration parameters, go back to the home screen and click the "Emergency Services Status" button. In the emergency status screen, click "Refresh Location Information and Service Mapping" to perform HELD and LoST queries. After a moment, you should have location information and a LoST mapping: The location information you provisioned into the LIS should appear in the "HELD Location Configuration" box, and the LoST information you supplied should appear as "Emergency Services Discovered by LoST". If you end up with location information but not LoST information, check to make sure that the mapping supplied in the LoST server is actually returned for the location in the LIS.

If the HELD and LoST queries succeeded, you should be ready to make an emergency call. To make the call, simply enter "911" in the field at the top of the window and press "Enter" or click "Go". This will trigger Zap! to recognize that you're placing an emergency call, which will automatically enable location conveyance. Click "Call" to make the call. When the call arrives at the PSAP PC (as it should instantaneously), it should display the caller's location. When the PSAP answers the call, the PSAP's Zap! client will show the caller's location at the bottom of the call window; click on "Location" to show it on a Google map. Congratulations, you've successfully placed and received an ECRIT emergency call!

References

Barnes M, Winterbottom J, Thomson M and Stark B (2010) HTTP Enabled Location Delivery (HELD). RFC 5985.

Geolocation API Specification (2010). World Wide Web Consortium.

Polk J, Schnizlein J and Linsner M (2004) Dynamic Host Configuration Protocol Option for Coordinate-based Location Configuration Information. RFC 3825.

Schulzrinne H (2006) Dynamic Host Configuration Protocol (DHCPv4 and DHCPv6) Option for Civic Addresses Configuration Information. RFC 4776.

Thomson M (2010) Using HTTP GET with HTTP-Enabled Location Delivery (HELD). Internet Draft (work in progress) draft-thomson-geopriv-held-get.

7

Security

The security of emergency calling systems is clearly a critical theme to address. Abuse of emergency calling systems can deny help to people in need – or even put them at greater risk than they were before – and waste valuable PSAP and first-responder resources. Each of the parties involved in an emergency call puts something valuable at risk by participating, and the goal of implementing security measures is to protect these things:

- Most obviously, the user is relying on the emergency calling system for his health and safety; by placing an emergency call, he is betting his own safety on the proper functioning of the emergency calling and response systems.
- PSAPs likewise undertake risk when they create an Internet-accessible interface, opening themselves up to all the current threats that Internet access poses (from malware to denial-of-service attacks), as well as the application-layer threat of false calls.
- Location providers (e.g., ISPs) support emergency calling by making location information available to callers, but at the same time, they may regard location information as a valuable asset.
- VoIP providers frequently charge for access to their network, but emergency calls are usually required to be free of charge, so there is a risk that false emergency calls will be used to obtain free calling services.

In this section, we will consider security mechanisms for emergency calling from three different perspectives: First, we will look at the foundational security features that ensure the proper functioning of ECRIT architecture. Second, we will consider how calling devices and location providers can create a system for protecting and assuring location information. Finally, we will outline a set of measures that PSAPs can

VoIP Emergency Calling: Foundations and Practice Karl Heinz Wolf and Richard Barnes

take to create an IP interface that is both open enough and safe enough to meet their needs.

Traditional PSTN-based models for emergency calling of course provide something of a "worked example" when it comes to security, since current systems are generally regarded as useful, stable, and secure. Indeed, the goal of many of the security mechanisms discussed below is to enable "PSTN-grade" security within new IP-based systems. However, it is important to keep in mind that creating the same assurances that were present in the PSTN might be more difficult or even impossible for IP-based networks. This trade-off is a basic consequence of moving from a vertically-integrated, locally-scoped system to a system that benefits from the layered model and location-independence of the Internet. Nonetheless, many of the security mechanisms discussed below can be thought of as a translation of trust relationships that already exist in the PSTN from one medium to another (e.g., from circuit connections to digital certificates) or from one type of entity to another (e.g., from telcos to ISPs).

7.1 ECRIT Security

The most fundamental requirement for all of the stakeholders in an emergency call is that the system should ensure that callers can be connected to the proper emergency response resources. Enabling this assurance is the goal of the security mechanisms specified in the ECRIT architecture. Recall that the three basic steps in the ECRIT emergency calling process are:

1. Determining the caller's location.
2. Determining the proper PSAP using LoST.
3. Delivering the call to the PSAP.

Security measures for each of these steps build on each other to provide assurance for the overall calling process, as illustrated in Figure 7.1.

7.1.1 Determining the Caller's Location

In the ECRIT architecture, the calling device gets location information from one of two places, either from some capability of the device itself (e.g., a GPS chip) or from a Location Server using some Location Configuration Protocol (LCP). Each of these options, and each different

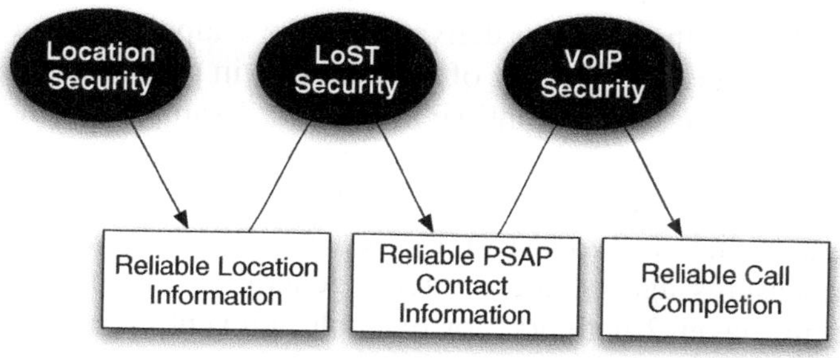

Figure 7.1 Building up a secure call by securing the steps in the ECRIT process.

LCP, have a slightly different set of risks and security mechanisms for the integrity of location information.

One might think that if a device determines its own location, then it can have absolute trust in that location. This is nearly true, but the modifier "absolute" is slightly overstating the situation, since even device-based positioning mechanisms can be subject to attack. For instance, there are well-known techniques for generating GPS signals that will cause a device to compute incorrect location values. Techniques that rely on an accurate compass can be misled by strong magnetic fields. Unfortunately, because these attacks are fundamentally physical, there are no real protocol mechanisms to counter them, beyond comparing results with another source (another device-provided measure or an LCP). At the same time, however, the physical nature of these attacks makes them difficult to mount compared to higher-level attacks, especially from far away or covering a large set of users.

As an aside, it is important for application developers to keep in mind that although applications running on a device might gain access to location through a single location API, this single API may provide location information from multiple sources. For example, several current devices use a single API to provide applications with access to GPS-derived location as well as location information acquired from a Location Server based on WiFi and cellular signals – without giving the application a way to distinguish between the two. Therefore, when analyzing the security of their applications, developers must consider the location functions underlying the location APIs they use.

The ECRIT architecture requires endpoints to support two LCPs (DHCP and HELD) and encourages the use of layer-2 mechanisms for location configuration. For both DHCP and layer-2 mechanisms, the reliability of the location delivered to the endpoint is essentially a function of the layer-2 security of the network, in particular with regard to integrity protection and anti-spoofing. For example, in the context of DHCP, an attacker can cause an endpoint to accept a false location either by modifying an existing DHCP message (e.g., as it passes through a malicious switch) or by acting as a rogue DHCP server. Countermeasures against these threats generally amount to maintaining a secure layer-2 network, for example, preventing endpoints from acting as rogue DHCP servers.

The security story for HELD is slightly more complicated, due to the fact that the overall process of getting location information via HELD has two steps: First the device discovers its local Location Server, then it sends a HELD query to that Location Server.

The primary security requirement for the discovery process is to ensure that the device discovers the proper Location Server URI. Recall that the discovery process involves getting a domain name from DHCP and resolving it to a Location Server URI using the DNS. So assuring that discovery provides the correct URI requires integrity protection for DHCP information and DNS information. For DHCP information, the situation is largely the same as above; the security of DHCP comes down to the security of the layer-2 network. For DNS, the most obvious mechanism to apply is DNSSEC (Arends et al., 2005), which allows the organizations that are responsible for parts of the DNS database to sign DNS records. At the time of this writing, however, DNSSEC is not widely deployed, but there are some mitigations available. Most importantly, the domain names that are used for Location Server discovery are *local*, in the sense that they refer to, and should be controlled by, the local network. Thus, for instance, an ISPs local resolver could be configured to treat these domains as "always in cache", which would prevent cache poisoning and assure that users get the current Location Server URI from the DNS (unless they use an alternative resolver).

Once the device has obtained a Location Server URI, the only remaining step to secure the HELD process is for the device to ensure that the Location Server that it connects to is in fact the server indicated in the URI. For this assurance, the device need only conduct the HELD transaction over HTTPS, in which case the standard name checking defined for HTTPS (Rescorla, 2000) (and implemented in standard HTTPS libraries) will ensure that the authenticated identity of the server matches the server name in the URI.

7.1.2 Determining the Proper PSAP

The process for discovering and querying a LoST server is much the same as the process for discovering and querying a HELD server. First, there is a discovery procedure based on DHCP and the DNS, followed by an HTTP query to obtain mapping data. So the security considerations are largely the same: Lower-layer security helps assure the DHCP information, DNSSEC helps with the DNS, and TLS provides authentication and integrity-protection for the LoST query over HTTP.

One important difference, though, is that unlike Location Servers, LoST servers are not expected to be operated by each access network, since they're not as intimately tied in with the physical network. Rather, LoST servers will likely be provided on a regional level, in order to provide information about that region, and maintain that information under the control of regional authorities. For instance, the emergency authority for a state might create a LoST server and invite local network operators to direct clients to it for local mapping information.

The security impact of this architecture is that the domain name under which LoST discovery records will be hosted will likely not be controlled by the caller's local access network. In order to fetch these records, either the caller or his local resolver will need to send a DNS query to an authoritative server outside the local network, which increases their exposure to DNS cache poisoning attacks. Thus, while HELD server discovery can get some security benefit from the locality of operations, LoST discovery must rely more on DNSSEC to secure its DNS transactions.

7.1.3 Delivering the Call

Next, we need to consider how the call itself is assured, that is, how the user can be assured that he is connected to the proper PSAP. The first question here is how the call reaches the PSAP. For this function, the user has to trust any intermediaries that process the call (e.g., SIP proxies). One interesting feature of the ECRIT architecture is that such intermediaries are not technically necessary. If the calling device has obtained a PSAP URI using LoST, then it can place a call directly to that URI. For example, if the PSAP URI is a SIP URI, then the calling device can simply send a SIP INVITE message directly to the PSAP URI.

This direct calling functionality comes with a few caveats, however. First, it needs to be supported by the underlying communications or

VoIP protocol. SIP supports direct calling (as described above), but XMPP does not, because it requires messages to be routed through intermediaries. And in practice, most VoIP endpoint implementations (including SIP, H.323, and others) require an outbound proxy. Second, direct calling needs to be supported by the PSAP, that is, the PSAP needs to have a policy that it will accept calls directly from endpoints, as opposed to, for example, from a white-listed set of providers. Finally, in practical terms, if an endpoint uses a different path for emergency calls than normal calls – for example, if emergency calls are direct while normal calls use an outbound proxy – then the code implementing emergency calling will probably not get the same level of testing and everyday use that the normal VoIP software sees, increasing the likelihood that bugs will go unnoticed until an emergency call.

In addition to getting assurance that the call is delivered to *someone*, the emergency calling system should also provide the user with assurance that the call has been delivered to the *correct* endpoint (i.e., the endpoint indicated by the PSAP URI). Mechanisms for providing this will again depend on the underlying communications protocol, but by the same token, existing security mechanisms can be applied to emergency calls just as they can be to normal calls. So, for instance, the SIP mechanism for authenticating the identity of the party to which a call has been delivered (Elwell, 2007) can likewise be used for an emergency services domain to authenticate a PSAP to the caller.

7.1.4 Considerations for Proxies

The ECRIT architecture allows for proxies to perform the routing steps discussed above on behalf of an endpoint. As discussed in Section 3.8, however, this model introduces significant fragility into the process. The primary security questions are related to discovery and authorization, namely how a routing proxy can discover a Location Server and a LoST server that can provide information about the caller, and whether the proxy is authorized to query those servers.

There are no good general answers to these questions. The discovery problem is easier to approach, since there are existing DNS-based discovery mechanisms for both LoST and HELD that can be redirected to allow third parties to discover the proper servers for an IP address. If the organizations that operate these servers publish the NAPTR discovery records in the reverse DNS tree, then proxies can perform a reverse DNS lookup on the client's IP address to find the proper servers.

Nonetheless, there is still some uncertainty with this method arising from the fact that the proxy cannot always tell whether the apparent source IP address of a call actually identifies the caller, as opposed to a NAT or proxy (in particular, Session Border Controllers frequently replace caller IP addresses with their own).

The security of these discovery techniques relies even more heavily on DNSSEC than do the basic discovery mechanisms. When an endpoint is discovering the local LoST or HELD server, it can get some assurance from the fact that there are mitigations against DNS spoofing on local networks (as discussed above). Third-party discovery cannot benefit from these mechanisms, and thus the only integrity protection available comes from DNSSEC.

The authorization problem is more straightforward for LoST, but harder for HELD. LoST information is generally regarded to be public information, so one does not anticipate problems with proxies being authorized to access LoST mappings. Location information, on the other hand, is very sensitive, so it is very likely that a Location Server will require an entity seeking precise information about an endpoint to have proper authorization. It should be noted, however, that proxies don't really need precise information about the caller's location, only a location that is precise enough to determine emergency call routing. Since emergency call routing is typically determined on a municipal, state, or national level, this level of location information is much less risky. As will be discussed in Section 7.2 (for different reasons), it makes sense for Location Servers to provide "routing-grade" location without requiring authorization, even if fine-grained location requires special authorization.

7.2 Location Security

In Section 7.1, we looked at one aspect of location security, namely how the caller's device is assured that the location it uses for emergency calling is correct, in the sense that it was obtained from the Location Server for the local network. In this section, we address a few other important security questions with respect to location:

Location Privacy: How the target of a location object can securely distribute that location object.

Location Assurance: How third parties (besides the target and the Location Server) can be assured of the authenticity of a location object.

Location Protection: How a location provider can provide enough location information to support emergency services without putting detailed information at risk.

It should be noted that none of the mechanisms described here or in Section 7.1 actually validate the *correctness* of location information; they do not assure that the location presented is actually the location of the client. In practice, that assurance is almost impossible to make (unless the target is physically incapable of moving!), and it certainly can't be proven through protocol mechanisms. Instead, we use authentication and integrity protections as a proxy for correctness: A location object is more trustworthy if the recipient can verify that it was received unmodified from a trusted source.

7.2.1 Location Privacy

It's important for certain entities to have access to the caller's location in the course of an emergency call – especially the PSAP – but the caller's location shouldn't be exposed to unauthorized entities in the course of an emergency call. Location privacy protections provide the access controls to manage access to location information. At base, privacy protections are a composite of authentication, authorization, and confidentiality protections: An entity that is granting access to location information must first verify the identity of the recipient, check that the recipient is authorized to receive location information, and protect the location information while it is en route to the recipient. This general pattern, though, has a few different concrete realizations, which we'll discuss below.

In this sense, the distinction between using location by value and by reference is important. As discussed in Section 3.2.2, using location by value means using a direct, intelligible representation of the location object, such as a PIDF-LO object, while in the context of location by reference, location information is represented indirectly, as a URI.

The clearest example of when a caller's location is at risk is when the caller transmits that location information in emergency call signaling messages, as described in Section 3.2.3. This obviously exposes the location information in the message to eavesdroppers between the caller and the PSAP. So in the interest of preserving the confidentiality of location, all emergency call signaling messages that contain location information should be carried over an encrypted channel.

Moreover, in VoIP protocols that use intermediaries, for example, SIP, location information can be exposed to intermediaries along the path between the caller and the PSAP. These intermediaries are of course authorized to inspect signaling messages, since they need information in them to route the messages, and sometimes they are authorized to inspect location information in these messages (e.g., to perform location-based call routing). In many cases, however, the location information in a message is intended only for the PSAP, which raises the question of how this location information can be protected from access by unauthorized intermediaries.

When location is transmitted by value inside a message, there are basically two options. If the caller has a public key for the PSAP, he can transmit location information in encrypted form, so that only the PSAP can access it. This situation, however, is exceedingly rare, because of the key management challenges, and most SIP implementations do not support encrypted message bodies. When a message contains unencrypted location information, the best that can be done is to advise intermediaries not to look at it. The SIP Geolocation header actually includes a special flag for this: The "routing-allowed" header parameter indicates whether SIP proxies are allowed to use the enclosed location to route the message.

Transmitting location by reference gives the target far more flexibility in protecting his location. In order to access the location information reference by a location URI, each recipient has to individually make a query. When the Location Server hosting the URI receives a query, it can authenticate the requester and apply any privacy rules installed by the user, as shown in Figure 7.2. So if the caller transmits a location URI in his SIP INVITE instead of a PIDF-LO object, then he can instruct the server hosting the URI to provide access only to the PSAP, not to any intermediate proxy. This might prevent proxies from performing location-based routing, but it is a very strong privacy control.

As described in Section 3.2.1, PIDF-LO location objects also carry some basic privacy rules directly in the location object. These rules inform the recipient of the location object of the user's privacy preferences, so that the recipient knows, for instance, how long he may retain the object.

There are two ways that a user can obtain location references. HELD includes a mechanism for provisioning endpoints with location URIs in addition to location values, so a user can get a location URI from HELD. But the user can also upload his location, by value, to another server, and use a reference to the uploaded location that server. Since

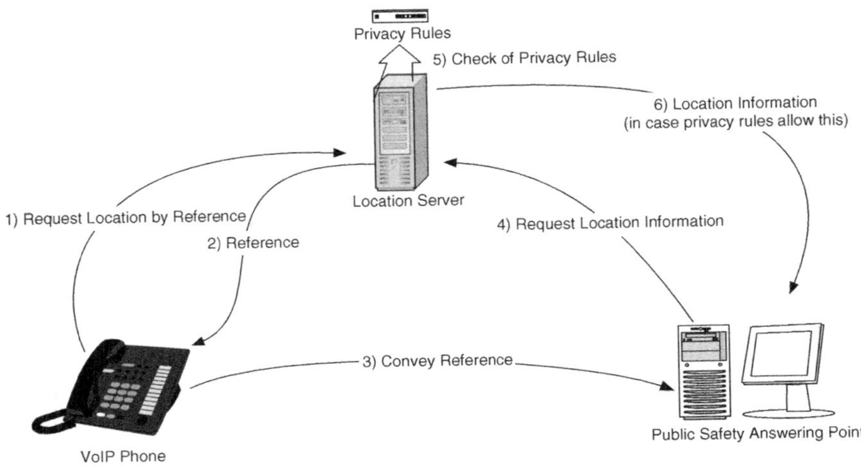

Figure 7.2 Location by reference with application of privacy rules.

this latter case is a close analogue presence service that distributes a user's presence information according to privacy rules, we will refer to a server where the user uploads location as a "presence server" to distinguish it from a general "Location Server". (This upload and redistribution can be accomplished via the SIP presence system (Peterson, 2005), but can also be done over other protocols. For instance, the Yahoo! FireEagle service provides such functionality over HTTP.)

The main difference between these two methods is whether access control is based on privacy rules or timing. Since the HELD protocol has no standard way for the user to communicate privacy rules to the server, the user has to accept the Location Server's default policy, that is, to be safe, the user has to assume that there is no access control on location URIs obtained through HELD. To mitigate this open-access policy, URIs distributed in HELD often have a prescribed "lifetime", after which all requests will be denied. (In addition, location URIs from a given network's HELD server are typically only valid while the target is connected to that network, so if a target is mobile, the validity period of location URIs from HELD is inherently limited.) In contrast, in the presence model, a user is typically assigned a single long-lived URI, but access to this URI is controlled with privacy rules.

In the SIP presence context, rules are typically expressed in the "common-policy" format (Schulzrinne et al., 2007) and managed using the XCAP protocol (Rosenberg, 2007). When managing geolocation information in particular, there is a geolocation-specific policy language that enables restrictions based on the target's current position and "fuzzing" to make location less precise (Schulzrinne et al., 2010). As of this writing, there is a proposal in the IETF GEOPRIV working group to allow rules from these languages to be installed on a HELD server to manage URIs distributed in HELD (Barnes et al., 2010), but no standards have yet been defined.

The primary challenge to the usability of privacy rules is how authorizations are managed, and in particular, how identities are shared and authenticated. A user may wish to specify a rule that says that only his local PSAP can access his location, but in order to do this he has to know the identity that the PSAP will use to authenticate to the Location Server, something that the standard mechanism for delivering information about PSAPs (LoST) has no way of telling him. Since it is difficult for the user to specify rules that grant PSAPs access to precise location, there is a risk that users will specify rules that apply globally (e.g., "Grant anyone access to my location, to the country level"), which will degrade PSAPs' access to location.

In this regard, URIs provided by local HELD servers have an advantage over presence URIs: Since the HELD server operated by a local network has an inherently local scope, the network operator that runs the HELD server can establish trust relationships with PSAPs in the region and issue them with credentials that they can use to authenticate. The HELD server can then employ a default policy that provides PSAPs with precise information, but provides anyone else with an imprecise "privacy-safe" location value. Establishing these relationships would be infeasible for a presence service that serves the Internet globally – or even one that serves a single nation like the US, which has more than 6,000 PSAPs.

Of course, location information is also exposed in the LoST protocol, but in this context, the location information cannot be associated with any other identity for the user than his current IP address. In contrast, location in a SIP message can be associated with the callers' SIP URI, a much longer-lived identifier. In addition, the LoST protocol is more "point-to-point" than SIP (in the sense that it runs over HTTP, which uses proxies far less commonly than SIP), so intermediaries pose less of a threat. So while it is recommended that LoST transactions be protected with TLS (for this reason as well as to authenticate the server), the protocol provides no further privacy protections.

7.2.2 Location Assurance

For PSAPs, it is important to know whether the location they receive with an emergency call is valid. Without any indication of the validity of location, a user can create a location object representing any location in the world, use it in a LoST query to find the local PSAP, then send a call to the PSAP carrying that location object. On the other hand, if a third party (e.g., a local Location Server known to the PSAP) vouches for the validity of the location, the PSAP can have some assurance that the caller is actually at that location. Moreover, if potential attackers know that PSAPs will have access to their precise location information (as a basis for follow-up action), this can act as a deterrent to false calls (Tschofenig et al., 2010).

The basic goal of location assurance is to provide a secure channel for location objects between a Location Server and a location recipient, for example, a PSAP. This channel should prevent modifications to the location object, and should enable the location recipient to authenticate the origin of the location object. This combination of properties assures the recipient that he has received the same object that the authenticated source meant to send, so if the PSAP views this source as trustworthy, then it can have more trust in the location object.

Such a secure channel can be established, in a manner of speaking, by value or by reference. That is, the Location Server can provide authenticated location information either by providing signed location objects or by providing location URIs (Marshall, 2010). With signed location objects, the security is provided by the object itself; the end recipient simply verifies the signature on the object without having to know how it arrived. This enables signed location objects to be handled in the same way as non-secure location objects, as shown in Figure 7.3. On the other hand, with location URIs, the security is provided by the channel over which location is delivered, using the standard process for delivering location by reference shown in Figure 7.2.

Both of these cases present problems related to key management and trust establishment. In either case, the recipient has to be able to authenticate the source of the location object. The main difference between these two approaches is the persistence of the assurance: If an entity receives a signed location object, he can pass it off to someone else, and the second recipient will still be able to validate the source of the location object. If the location object is delivered over a secure channel in response to a query for a location URI, the recipient cannot retransmit the location to someone else and still maintain the assurance. The

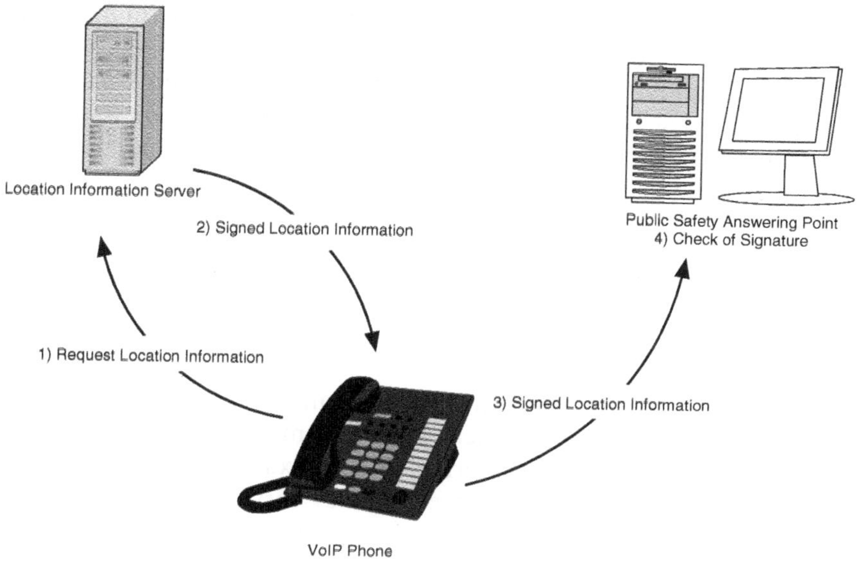

Figure 7.3 The Location Server signs a location object, and the PSAP verifies the signature.

only way the secondary recipient can be assured of the location object's source is if the first recipient gives him the URI and he downloads the object himself. Since the second recipient may not be authorized to download the location object, he may not in fact be able to validate the authenticity of an object that is given to him. So location URIs have the curious property that they can be used not only to restrict access to location information, but also to restrict access to location *validation* information – if you can't access it, you can't validate it.

Depending on the Location Server, using location URIs for security can still expose location recipients to some risk of spoofing. In particular, if a Location Server allows users to upload location information to be made available through a URI (as is often true, e.g., in the SIP presence model), then a malicious user may be able to get the presence server to vouch for a spoofed location. Whether or not this is possible will depend on the Location Server, so PSAPs and other location recipients should take into account policies about user-provisioned location when deciding whether or not to trust a given Location Server.

Finally, note that if an end device determines its own location (via GPS or directly from the user), none of these mechanisms really apply, since there's no third party who is authoritative for that location object.

So as long as there are devices out there that use self-determined locations for emergency calling, PSAPs will have to accept calls without validated location (either signed or from an authenticated URI). This is not to say that location assurance techniques are without value, but rather that they should be used as a weighting measure, and alongside other techniques.

7.2.3 Location Protection

Many location providers view location as a valuable resource, so that if location information is going to be used for commercial services, they would like to be paid. Given that location information is required for emergency services, and emergency services must be free of charge in most jurisdictions, the question arises of what exactly location providers must provide in order to support emergency services. (This subject is also covered in an ECRIT Internet-draft (Barnes and Lepinski, 2010), which we summarize here.)

Keeping in mind that the two major uses for location in emergency calling are call routing and dispatch of first responders, the general answer to this question is that the location provider must provide the entity that performs call routing with location information that is usable for routing, and it must provide the PSAP with location information that is usable for dispatch. Note that since the Location Server doesn't know who the routing entity is, it will need to provide routing location to any requester. So we have two notions of "usability" to define, which will clearly have very different definitions.

Providing location that is usable for dispatch generally means that the location provider will make available to the PSAP the highest precision that it has available. If a less precise location is to be given to others, then this requires that PSAPs be authenticated. As discussed above, however, maintaining trust relationships with local PSAPs is not a severe burden on the location provider, since any given Location Server will only have to deal with a few PSAPs (not the entire world).

A "rough location" that represents an approximation of a precise location must meet three criteria in order to be usable for routing:

1. A LoST query with the rough location and any Service URN must result in a single mapping.

2. This mapping must be the same as the one returned for the precise location.
3. The rough location must contain the precise location as a geographical subset.

That is, emergency call routing with the rough location must work in exactly the same way as routing with the precise location.

Based on these criteria for rough location, we can lay out a procedure for determining the largest possible area (i.e., the roughest possible location) that still meets the criteria; any other rough location must fit within this region. Recall that each LoST mapping comes with a "service boundary" within which the mapping is valid (the service boundary is the boundary of a given PSAP's jurisdiction). For each service, rule (2) above requires that the rough location be within the same service boundary as the precise location, while rule (1) requires that it should not overlap with any other service boundaries. The set of points that meet these criteria is exactly the service boundary that contains the precise location!

Since the rough location has to be contained in the service boundary for each service, the largest possible rough representation of a precise location is the intersection of all service boundaries that cover the precise location. This means that all the Location Server has to do to compute the rough location for a point is do one LoST mapping query per service, then take the intersection of all the service boundaries. The server can then return any polygon that contains the precise location, that is, any location value that meets the third rule. This process is illustrated in Figure 7.4.

Finally, in order to allow different entities access to different grades of location, it is important for a Location Server that provides rough location to also provide endpoints with a location URI that authorized entities can use to access precise location. Calling devices, in turn, need to relay both the rough location by value and the location URI. Thus, if a caller connects to a network with such a server, he will receive both a location object and a location URI, and will include both of these in his emergency call signaling. When the PSAP gets the signaling message, it can dereference the location URI to obtain the precise location.

As we have discussed elsewhere, there are other reasons for Location Servers to have this "dual-precision" policy as the default for location URIs they issue, that is, to provide a rough location to any requester

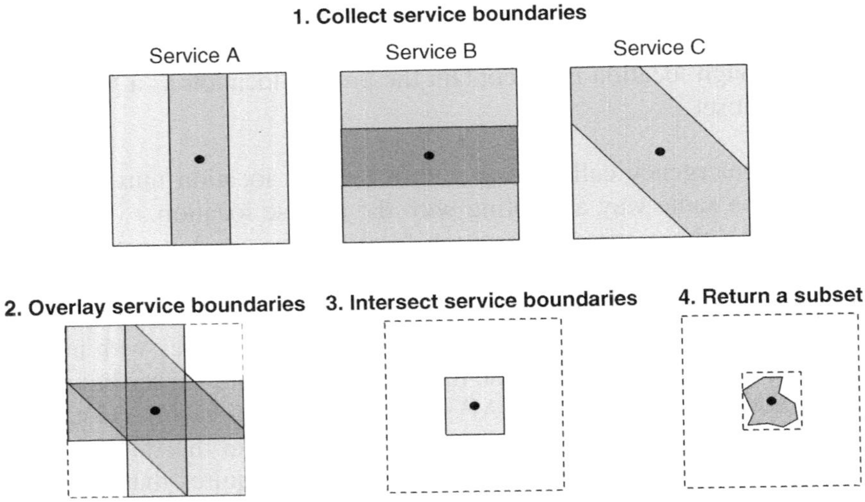

Figure 7.4 Computing a rough location from a precise location.

and a precise location only to authorized ones. For example, such a policy prevents proxies from gaining access to precise information about the caller's location, but provides them with enough information to route an emergency call, minimizing the privacy risks of routing a call through proxies.

7.3 PSAP and VoIP Network Security

In accepting emergency calls from IP endpoints, both PSAPs and VoIP providers are increasing their exposure to fraud. For VoIP providers, the risk is that callers will exploit the fact that emergency calling is provided free of charge in order to get free calling (i.e., free *non*-emergency calls). For PSAP, the risk is that they will be accessible from anywhere in the Internet, so that attackers from anywhere can send them calls, possibly overwhelming the resources available at the PSAPs.

Put differently, VoIP networks and PSAPs both require certain assurances about putative emergency calls. VoIP networks need to know that a call that looks like an emergency call is actually addressed to a PSAP (as opposed to some non-emergency destination). PSAPs need some assurance that a call is actually related to an emergency, and that the caller is actually within their jurisdiction.

7.3.1 Basic PSAP Protection Measures

Protecting PSAPs is perhaps the most important security goal for IP-based emergency calling. If an attacker is able to disable a PSAP, then he can effectively deny emergency services to an entire region or perhaps an entire country. In current, PSTN-based emergency calling systems, PSAPs are mostly protected by the security of the PSTN signaling plane, in that calls can only be directed to them by authorized providers, along a fixed set of circuits. (Even still, there are attacks against this system, e.g., through caller-ID spoofing.) Simply connecting a PSAP to the Internet provides none of this protection; creating secure IP interfaces for PSAPs will require adding mechanisms to create the necessary protections.

Ignoring for the moment PSAPs' role in answering emergency calls, PSAPs have many characteristics in common with other high-availability Internet services, for example, large-scale commercial websites or corporate data centers. Many techniques have evolved over time to ensure the reliability of these services, which can now be applied to PSAPs.

Standard boundary security techniques such as firewalls and intrusion detection systems can of course be used as an initial barrier to unauthorized access from the Internet. Some firewalls may need to be specially configured to make sure SIP calls can be received reliably, but actually, the specialized nature of PSAP could allow for tighter boundary control than in other cases (such as enterprise networks) where many different services need to be allowed through. The NENA i3 specification (*NENA Functional and Interface Standards for Next Generation 9-1-1 Version 1.0 (i3)*, 2007) is strongly focused on this style of protection, establishing an Emergency Services IP network (ESInet) with a secure boundary, within which PSAPs and emergency services can freely communicate.

Redundancy and fail-over are important considerations as well, in order to deal with overflow or denial-of-service situations. These concepts should be implemented at many layers. To be reachable anywhere over IP, PSAPs need to have reliable IP connectivity. In order to assure connectivity, it can be useful for PSAPs to maintain multiple connections to the Internet, from different providers (the technical term is "to be multi-homed"), so that if one connection goes down, traffic can be re-routed over the others. One current example of this technique is found in Sweden, where current IP-connected PSAPs are connected to at least four different Internet exchange points, spread all across the country. Connecting to an exchange point also gives PSAPs the opportunity to connect to multiple different upstream ISPs for connectivity to the rest of the Internet.

LoST provides an important tool for redundancy and failure as well, since a LoST server directs clients to a particular PSAP. For example, if one PSAP fails, then changing the LoST mapping for that PSAP so that it includes a PSAP URI for a backup PSAP (e.g., in a neighboring city) will cause all emergency calls for the failed PSAP to go to the backup PSAP. LoST mappings can also provide multiple URIs to enable client-side load-balancing, as is commonly done with mail server (MX) records on the DNS.

Of course, the LoST service itself should also be redundant, as well as the location services and DNS services that support LoST discovery. Standard techniques for service redundancy apply here as well. In particular, there should be multiple name servers, Location Servers, and LoST servers in diverse geographical locations, with diverse connections to the Internet. Clients can be instructed to use all of these different servers by including multiple name server (NS) records or LoST/HELD discovery NAPTR records in the DNS.

7.3.2 PSAP Fraud Mitigation

Beyond the basic threats posed by Internet connectivity, PSAPs also expose themselves to a set of application-layer risks by accepting VoIP calls. The basic ECRIT architecture defines how an endpoint can deliver a call to the proper PSAP for a given location; this same architecture, however, can be used by an attacker to determine the proper PSAP to target with fraudulent calls, enabling him to overwhelm the PSAP's ability to handle legitimate calls.

It's worth noting that this risk is not so different from the risk of hoax calls that PSAPs currently face. The additional risk introduced by VoIP is that there are no longer constraints in the calling network that help assure that the caller is within the PSAP's jurisdiction. So a major security goal for VoIP 9-1-1 will be to re-create these constraints in new ways.

The basic criterion for a call to be valid is that it must relate to an ongoing emergency within the PSAP's jurisdiction. While the PSAP might be able to forward calls for an emergency outside of its authority to the proper destination, this is not generally the primary purpose of the PSAP. And calls that aren't related to an emergency at all should not consume the PSAP's valuable resources. Unfortunately, automated security measures cannot definitively tell PSAPs which calls are valid and which are not, but they can provide some signals that can be used, for example, to prioritize incoming calls. In this regard, such mechanisms are similar to spam detection mechanisms for email, in that they will not define a hard cut-off, but rather a "score" of how good or bad a given call is.

Of course, PSAPs should perform some basic checks that an INVITE message is well-formed. It should be addressed to the PSAP, in the sense that the ECRIT architecture defines: The Request URI should be set to a Service URN that identifies a service for which the PSAP is responsible; the Route header should be set to the PSAP's contact URI; and the Geolocation header should specify a location that is within the PSAP's jurisdiction. The failure of any of these checks represents a basic fault in the emergency call routing system, and should be flagged to the call taker that answers the call.

Although it is really infeasible to tell with automated mechanisms whether the *subject* of an emergency call is related to the PSAP's jurisdiction, there are a few mechanisms to help assure that the *caller* that placed an emergency call is within the PSAP's jurisdiction. Most directly, the location assurance mechanisms described in Section 7.2.2 (and in the ECRIT document (Tschofenig et al., 2010)) can be used by location providers within the PSAP's jurisdiction to vouch for location they provide.

When a PSAP receives a call with secure location information (either a signed location object or a secure location URI), it can immediately check whether the location information is provided by a trusted source. But even if location does come from a trusted source, the PSAP will still need to verify the binding between the location and the call, in order to prevent attackers from re-using validated locations for calls that are distant (in time or space) from the validated location object. At a minimum, the PSAP can validate that the timestamp on the location object is acceptably close to the time at which the call was received (to prevent an attacker from re-using an old location object). It is more difficult to bind the location object to the identity of the caller (to prevent two callers trading location information), especially given that Location Servers in general will not be aware of the application-layer identities of clients.

However, the "entity" field in a PIDF-LO can be used to convey identity information for the target, so a PSAP could work with location providers in its region to agree on the contents of this field. For instance, location providers could assign entity URIs that encode the IP address of the caller, which the PSAP could match against the source address of the call (although this would be broken by some SIP intermediaries). A network that controls both positioning and calling (e.g., an IMS network) could actually set the "entity" field to the caller's SIP URI.

Another way for PSAPs to have some validation of the source of calls is to use the fact that the IP routing infrastructure has some locality properties. If the PSAP collects information from local ISP on how they assign IP addresses, then in some cases, it can determine which blocks of IP addresses are in use within its jurisdiction. Then, when a call comes in from a given IP address, the PSAP can determine whether it belongs to one of the blocks in use locally.

This technique, however, is only as reliable as the IP address that the PSAP has for the caller. When callers send calls directly to the PSAP, instead of through an intermediate VoIP provider, the PSAP can have fairly direct information about the caller's IP address – it is the source address of the incoming packet with the SIP INVITE. When a call goes through a VoIP service, however, even fields that are supposed to indicate the caller's IP address (e.g., media destinations in an SDP message) can be replaced with the address of an SBC or other intermediary. So while it can be tempting to use IP address validation to make a concrete determination as to whether a call is local (some have even suggested that PSAPs only accept calls over VPNs to local networks), it is much safer to use this technique as another signal rather than a hard discriminator.

Automatically validating that a call is related to a real emergency is even more impossible than validating the location of the emergency. In the PSTN, anti-fraud mechanisms rely in large part on the system's ability to provide PSAPs with accurate information about the caller's identity and location. This information enables local law-enforcement authorities to take action against fraudulent callers, and thus acts as a deterrent to false calls. The same approach will provide similar protections for IP emergency calls, except that the sources of this information might be different, since ISPs as well as VoIP providers will need to be involved.

In the PSTN, the same entity provides both the caller's physical connectivity and voice service, and most callers are required to have a subscription. This integration allows a single entity to provide information on the physical and application-layer identities of the caller (i.e., phone numbers), the caller's geographic location, and the caller's real-world identity (i.e., his name), as well as other associated information such as a home address.

In the context of VoIP calling, this information is more diffuse. The ISP serving a caller knows his IP address and any lower-layer identifiers (e.g., a MAC address or IMSI), but may not have access to an application layer identifier like a SIP URI. The caller's local ISP is also the entity that is responsible for locating the caller in the ECRIT architecture, so it is expected to have geolocation information. The caller's VoIP provider, by contrast, is responsible for the mapping between the caller's application-layer identity and his IP address, but may have no information about the caller's current location.

Either the ISP or the VSP might maintain subscription records for users, but neither is required to. For example, if a user connects anonymously to an open WiFi hot-spot, then uses a generic VoIP service that requires no subscription, then neither his ISP nor his VoIP provider will know his true identity. In fact, since callers can send calls directly to PSAPs, no VoIP provider is strictly necessary.

So there is a real challenge for PSAPs to collect actionable information about caller identity and location in this environment. PSAPs will need to maintain a broader set of relationships than they have in the past. In addition to local telecommunications operators, PSAPs will need to have relationships with local ISPs and possibly third-party location providers. One goal of these interactions could be to establish credentials for the PSAP to log into Location Servers, so that it can be automatically authenticated and authorized to retrieve high-precision location. No Internet-standard mechanisms currently exist for querying subscriber information, but some other protocols developed in the context of the PSTN could be useful, for example the Parlay/X suite of protocols (*Open Service Access (OSA); Parlay X Web Services; Part 1: Common*, 2009).

Since an IP-connected PSAP can in principle receive calls from a VoIP provider anywhere in the world (even though the caller should still be in the PSAP's jurisdiction), it is in general infeasible for a PSAP to maintain relationships with all VoIP providers that might send calls to it. Where these relationships can be established, however, they can still be useful. Many VoIP subscribers right now are covered by "PSTN-replacement" VoIP, in which the VoIP provider is the same as the ISP; like ISPs, then, these VoIP providers have a defined service area, and in some areas, they are already required to interconnect with local PSAPs. Beyond these inherently local services, PSAPs might also try to establish relationships with a few key "nomadic" or "pure VoIP" providers. Which providers will be the most relevant will depend on the PSAP; it will likely be beneficial for PSAPs to monitor which providers send them the most calls, then contact those providers.

These relationships between the PSAP on the receiving end of a call, and ISPs and VoIP providers on the originating end of a call can of course be used as another way to establish the trustworthiness of a call. A PSAP might regard a call as more trustworthy if it comes from an ISP that can provide location information or from a VoIP provider that can provide subscriber information. As with the other criteria for evaluating calls discussed above, however, matching against a list of approved ISPs and VoIP providers is best used as one input among many for predicting the validity of a call, rather than a basis for a hard decision to accept or reject a call. Filtering based on the VoIP provider delivering a call can also inhibit competition and innovation by raising the barrier to entry for new VoIP services or other communications media.

In all these cases, as PSAPs and regulators develop policies for filtering or prioritizing calls that arrive at a PSAP, they should keep in mind the trade-offs that such policies entail. On the one hand, permissive policies can increase the risk of a PSAP being overwhelmed by false calls, putting the entire community at risk. On the other hand, restrictive policies will prevent certain classes of users from accessing the emergency calling system. For example, if emergency calls are only accepted from VoIP carriers inside of a country, users of foreign VoIP service – whether visiting or native to the country in question – will be unable to place emergency calls. Filtering based on the source ISP can likewise lead to problems for users whose calls are routed – often without their knowledge – through a remote SBC.

7.3.3 VoIP Provider Call Validation

The main risk that VoIP networks face is also fraudulent calling, but in a different sense, namely that emergency call marking could be used to circumvent normal access controls, such as charging systems. Of course, this risk is only present if the VoIP provider is somehow limiting access in the first place on a call-by-call basis; providers that allow subscribers unlimited calling for a flat fee would have no access control concerns that would require distinguishing emergency calls from non-emergency calls (although they might still want to make this distinction, e.g., to assign priority to emergency calls). Nonetheless, many VoIP providers do still control access on a per-call basis (e.g., metering by call minutes and destination), so it's important for these providers to have a way of knowing when an call is an emergency call.

Consider the possible forms that an emergency call can take when entering a provider network, in terms of the SIP INVITE message that initiates the call. As described in Chapter 3, there are three fields in an INVITE message that are set to defined values for an emergency call:

Request URI: Set to the Service URN for the emergency service being requested.

Route header field: Set to the PSAP URI for the destination PSAP.

Geolocation header field: Set to point to the caller's location.

Actually, since we're talking about security from the provider's point of view, we should note that these are the values that *should* be filled in these fields; to validate that a call is an emergency call, the provider needs to validate the contents of these fields. Fraudulent calls might be marked with geolocation and a Service URN, but be routed to a destination other than a PSAP.

Note that when a call arrives at the provider, it may not have all three fields populated, since the endpoint might not have completed the full ECRIT emergency call routing process. For example, if the endpoint is not location-enabled, then the Geolocation header will be missing, or if the endpoint supports geolocation but not LoST, there may be values in the Geolocation and Request URI fields, but no PSAP URI in the Route field. If the endpoint does not support any part of the ECRIT process, then the call may simply arrive with a Request URI containing a "tel:" URI with the dialed emergency number, and no Route or Geolocation header fields at all. In all of these cases, the endpoint requires assistance from the VoIP provider in routing the call (as discussed in Section 3.8). Since it is the VoIP provider that is doing the routing in these cases, the provider has direct assurance that the call is going to a PSAP.

So the only case where there is a risk of fraud is from fully ECRIT-enabled endpoints that perform both location and call routing on their own. These endpoints will deliver a call to their VoIP provider with all three of the above fields pre-populated; they only rely on the VoIP provider to deliver the call to the specified PSAP. In this case, if the VoIP provider wants to validate that a call is an emergency call, then he needs to check the validity of these three fields.

This validity check is really a check of relationships among the three fields. Since the Geolocation field and the Request URI can in general contain any geolocation data and any Service URN (depending on where the user is and what service he desires, both of which are unknown to the VoIP provider), the VoIP provider cannot verify them directly. (For Geolocation, the provider may be able to obtain some validation via the location assurance mechanisms described in Section 7.2.3, but this would only validate that the location describes the end-points, current location, saying nothing about whether the call is an emergency call.) Likewise, since it is infeasible for a VoIP provider to maintain a global list of all PSAP URIs, it cannot check the Route header independent of other fields.

If the fields are populated as specified for an emergency call, however, they will have certain relationships that can be checked by a third party VoIP provider. In particular, the Service URN in the Request URI must represent a service that is available at the indicated location, and the PSAP URI in the Route header field must identify the PSAP for that service in that location. Both of these relationships can be validated with LoST queries. So the overall procedure for validating that a call is an emergency call is as follows:

1. If location is provided by reference, download the referenced location object.
2. Copy the location information and the Service URN from the message into a LoST "findService" query.
3. Send the query to a LoST resolver and collect any mappings returned.
4. Match the URIs in Route header fields in the message against URIs in the "uri" elements of the mappings.

If any of the Route headers in the message matches one of the PSAP URIs carried in the "uri" elements of the mapping, then the call is genuinely an emergency call addressed to a real PSAP.

VoIP providers should keep in mind, however, that this validation technique can produce "false negatives", since there are ways that the validation can fail even for valid emergency calls.

Note that all of these checks require that there be enough location information in the call that the provider can access, and that this location information be precise enough to support LoST queries. If location is provided by value, this is not a problem, since the provider can simply read the PIDF-LO object enclosed in the INVITE (although this may violate some of the user's expressed privacy preferences encoded in the "routing-allowed" header). If location is provided by reference, however, it is possible that the VoIP provider will not be authorized to access the referenced location information, and thus unable to validate that the call is an emergency call. Ensuring that calls will be able to transit VoIP providers is thus another reason for Location Servers that host location URIs to provide low-precision "routing-grade" location in response to any query, and for endpoints to include this rough location in emergency calls (alongside the reference to more precise location).

A more subtle failure mode involves the LoST infrastructure: If the LoST system returns different results to the VoIP provider than it provided to the caller, then the validation will fail. The most obvious case where this could happen is when the caller has access to LoST mappings via a local LoST resolver that is not accessible to the rest of the world, in particular, the VoIP provider. But there may also be cases where an emergency authority chooses to advertise different URIs to different queries, either because the URI has changed between the time the call was routed and the time the provider verifies it, or because there are actually multiple valid URIs (e.g., for load balancing).

These failure modes suggest that while this LoST-based validation can be a powerful tool, it should be handled carefully. Namely, VoIP providers should not set a policy to immediately drop calls that are not validated; rather, they should consider letting these calls proceed, but with controls, for example, lower priority or limits on the rate at which such calls can be made. In this sense, provider-based validation of calls will not fully prevent users from placing fraudulent calls using emergency call marking, but will at least limit the scope of such fraud.

References

Arends R, Austein R, Larson M, Massey D and Rose S (2005) DNS Security Introduction and Requirements. RFC 4033.

Barnes R and Lepinski M (2010) Using Imprecise Location for Emergency Context Resolution. Internet Draft (work in progress) draft-barnes-ecrit-rough-loc.

Barnes R, Thomson M and Winterbottom J (2010) Location Configuration Extensions for Policy Management. Internet Draft (work in progress) draft-barnes-geopriv-policy-uri.

Elwell J (2007) Connected Identity in the Session Initiation Protocol (SIP). RFC 4916.

Marshall R (2010) Requirements for a Location-by-Reference Mechanism. RFC 5808.

NENA Functional and Interface Standards for Next Generation 9-1-1 Version 1.0 (i3) (2007). Technical report, National Emergency Numbering Association (US).

Open Service Access (OSA); Parlay X web services; Part 1: Common (2009). Technical report, 3GPP.

Peterson J (2005) A Presence Architecture for the Distribution of GEOPRIV Location Objects. RFC 4079.

Rescorla E (2000) HTTP Over TLS. RFC 2818.

Rosenberg J (2007) The Extensible Markup Language (XML) Configuration Access Protocol (XCAP). RFC 4825.

Schulzrinne H, Tschofenig H, Morris J, Cuellar J and Polk J (2010) Geolocation Policy: A Document Format for Expressing Privacy Preferences for Location Information. Internet Draft (work in progress) draft-ietf-geopriv-policy.

Schulzrinne H, Tschofenig H, Morris J, Cuellar J, Polk J and Rosenberg J (2007) Common Policy: A Document Format for Expressing Privacy Preferences. RFC 4745.

Tschofenig H, Schulzrinne H and Aboba B (2010) Trustworthy Location Information. Internet Draft (work in progress) draft-ietf-ecrit-trustworthy-location.

8

Ongoing Emergency Calling Work

Although the ECRIT architecture for VoIP-based emergency calls is basically complete, some details are still under development and a few open issues remain. In addition to ongoing standardization and technical work, there is also a continuing debate about which entities will have to implement which parts of the ECRIT architecture, as discussed in Section 5.1. The sections below discuss ongoing standardization and prototyping work, open issues in the standards community, and implementation challenges – and some of the work that is being done to address these challenges.

Updated information on these issues will be posted on the book's companion website at `http://www.voip-sos.net/`. Check out this website to stay informed about ongoing issues and new developments.

Many of these standardization and implementation issues are being worked on in organizations already mentioned above, such as the IETF and 3GPP. One important group that we have not discussed yet is the Emergency Services Workshop (ESW), not so much an organization as a series of meetings. Initially started by a group of participants in 3GPP and IETF, the ESW series seeks to bring together the many stakeholders in next-generation emergency calling, including standards organizations, equipment vendors, network operators, regulators, and others. The hope is that these meetings can help the different systems being developed by all these different entities operate as a single global emergency calling system, and to help regulators make informed decisions about how to deal with these systems. ESW meetings are held roughly every six months; more information is available on the ESW website at `http://www.emergency-services-coordination.info/`.

8.1 Prototyping, Implementation, and Interoperability

Clearly, before VoIP emergency calling systems can be deployed in real networks and PSAPs, there need to be software and hardware products available that can implement the relevant standards. As discussed in Chapter 6, there are already several prototyping projects that implement different aspects of the ECRIT architecture, including location configuration, LoST mapping, and the placing of emergency calls. In addition to these projects, there are some other efforts underway to help ensure that implementations are interoperable, and to provide feedback on further standards discussions.

In the US, the most high-profile implementation and interoperability effort is a series of Industry Collaboration Events, or ICE Events, organized by NENA. The goal of these events is to bring together vendors of emergency services equipment to test interoperability among their devices. In organizing these events, NENA has had to balance the demands of end-users and PSAPs for information on product compatibility with vendors, need to manage information on their products' capabilities. To try to optimize this balance, the ICE events are thus conducted in a semi-confidential environment, where participants agree both to keep certain aspects of the test confidential and to open relevant results to the public.

Each ICE event is focused on a particular aspect of the US emergency calling system. The focus of the first event (ICE 1) was to "Test Location, Routing, and Delivery of NG9-1-1 Calls" (recall that NG9-1-1 is the NENA name for IP-based emergency calling). Future events will focus more on particular components: ICE 2 is devoted to "Transitional Elements" (i.e., the transition from PSTN to IP calling), ICE 3 is devoted to "Location-Related Issues", et cetera. Vendor interest in these events has been strong. ICE 1 attracted 16 different vendors, who built a testbed that incorporated 20 different NG9-1-1 components, and NENA is trying to increase these numbers for future events. More information about the ICE events can be found on `http://www.nena.org/ng9-1-1/ICE`.

The ICE events are largely modeled on the SIPit events organized by the SIP forum to test interoperability among SIP implementations. These events are held roughly every six months, in locations around the world, and typically involve around fifty different implementations. Like the ICE events, the goal of the SIPit events is to gauge interoperability and provide feedback to the standards community, while providing some confidentiality to participants; as the SIPit website notes, "while there is no formal non-disclosure agreement, the spirit of the event is to keep the details of testing confidential." The official website for the SIPit event series is located at `http://www.sipit.net/`.

The European Commission has funded several projects related to next generation emergency services. The one most closely related to ECRIT is the EU PEACE program, which involves research on "IP-Based Emergency Applications and Services for Next Generation Networks". (Here, the term "Next Generation Network" basically means the 3GPP IP Multimedia Subsystem, or IMS.) Participants in the EU PEACE program are developing prototype software for different parts of the emergency calling system (such as the Emergency branch of the Fraunhofer Open IMS Core described in Chapter 6), and integrating them into overall emergency calling and emergency management solutions.

There has actually been some productive interoperability testing at ESW and IETF meetings. Since many people interested in VoIP emergency calling are already at these meetings, they sometimes gather to test their implementations is a real network – the IETF's meeting network. With the help of the IETF network operations center, GEOPRIV participants have created HELD servers at the past few IETF meetings, which use the IETF 802.11 network to provide participants with location configuration services. In more than one of these tests, bugs have been found in both implementations and specifications, leading to improvements in both. Even better, feedback on the specification documents could be directly provided during the meetings.

Finally, there are several ongoing prototyping activities at nic.at, including the following (among others):

- further enhancement of the Zap! SIP client for placing emergency calls;
- Wireshark extensions related to location configuration;
- development of a LoST forest guide.

Above all, prototype implementations for VoIP-based emergency calling show that effective VoIP emergency calling is feasible. Hopefully, all of these continued prototyping and interoperability efforts will help with the deployment of the ECRIT emergency calling framework around the world.

8.2 Ongoing Standardization Issues

8.2.1 Default PSAPs

When all means to acquire location information fail (and there's not even a default location, as explained in Section 4.5), it is impossible to find out the responsible PSAP for a call. In order for emergency calling to still be possible in such situations, some standards documents define the notion of a "default PSAP". This notion is fraught with issues, however. First, it is not clear how such a default PSAP should be chosen, especially for calling devices, such as soft phones, which can place calls from anywhere on earth. Second, there would also need to be a default emergency dial string, or else there would be no way to recognize that an emergency call has been placed. The use of default PSAPs only really makes sense when a caller can only connect to a network in a geographically constrained region, which has a single set of emergency numbers and a single PSAP.

8.2.2 Unauthenticated Emergency Calls

An important question when developing an emergency call system for the Internet is whether unauthenticated calls should be accepted or not. Is a subscription with a VoIP provider necessary and must the phone be configured to use this provider, with login information correctly typed in? Or should an emergency call via the Internet be possible without any user configuration – and therefore without any authentication? The layered architecture of the Internet also introduces another issue: Access to networks can typically be used only by registered subscribers. Should there be a way to get anonymous access to the network in order to be able to place an emergency call (similar to using a foreign cellular network for an emergency call)?

The IETF ECRIT working group has done some initial work on this topic, mainly toward clarifying the nature of the problem (which is two-fold; gaining access to an access network, and placing the call). This problem statement is contained in an ECRIT document entitled "Extensions to the Emergency Services Architecture for Dealing with Unauthenticated and Unauthorized Devices" (Schulzrinne et al., 2010), but there are still issues that the document does not address. For example, the document suggests that a phone should allow unauthenticated emergency calls only when the jurisdiction requires that, which raises the following question of how a phone should know whether the jurisdiction for the phone's current location requires that unauthenticated emergency calls must work. The IETF documents do not give information on that. This information could be provided, for example, by a mapping server that could provide information on unauthenticated emergency calling requirements for a given location. Anyway, debate on this issue continues.

8.2.3 VPN Problems

VPNs or other tunnels can interfere with the location configuration process, when positioning is based on the endpoint's network attachment (e.g., based on its IP address). In the worst case, an endpoint performing location configuration through an established VPN tunnel could receive the completely wrong location information to the client. The only general solution so far (apart from using GPS receivers for location determination) is to perform location configuration before establishing the VPN tunnel. This solution is not complete, however, since VPN tunnels cannot necessarily be detected in every situation (e.g., inter-site VPNs). Hence, entities that operate VPNs and Location Servers need to make sure that Location Servers are aware of tunneling and take it into account. Even still, given the risk of incorrectly configured VPN operators, there is some residual risk that endpoints will get an incorrect location due to VPNs denying them the ability to make emergency calls via the Internet.

8.2.4 Home Emergency Dial String Issues

Some current emergency calling implementations that can roam internationally (e.g., GSM mobile phones) allow the user to initiate an emergency call both by entering the dial string for his current location, as well as by entering the dial string for the phone's home country (a "home dial string"). While this can be an appealing feature, reducing the amount of information the user has to recall during an emergency, there is some risk of misdirected calls. In particular, emergency dial strings for one country can be legal, non-emergency telephone numbers in another, so a user that dials a home dial string in such a situation might be directed to another emergency service, or a destination completely unrelated to emergency calling. For example, in Italy, 118 is the emergency number for the ambulance service, while in Liechtenstein it is the fire brigade, and in Finland it is just a directory information service. At present, there is no global plan for resolving such conflicts. While this issue isn't directly related to IP emergency calling, it does emphasize the importance of discovering local emergency numbers (see also Section 3.5.1).

There can also be differences in the way dial strings are used in different jurisdictions that introduce ambiguity with regard to where an emergency call should be routed when the user dials a home emergency dial string. For example, consider the case where the user has separate home emergency dial strings for police, ambulance and fire brigade configured. At the visited location no such differentiated services exist, with only a single PSAP taking all calls. So if any of the home emergency dial strings is dialed, the call has to be routed to this one PSAP.

There is also the reverse case: The user has a single emergency dial string but at the visited location only separate and specialized emergency services are available. In this circumstance, where should a call to the single home dial string be directed? In many jurisdictions, one emergency service can act as a single entry point to route calls to other services. For example, in Austria, the police can dispatch calls to the fire and ambulance services, but not vice versa. These relationships between services vary significantly between jurisdictions, so there's no way of knowing whether a given emergency service can be forwarded to another. In any case, there is no current specification that defines how these situations are handled.

8.2.5 Updating the List of Available Emergency Services – the LoST Service List Boundary

As already mentioned, PSAPs (as well as non-emergency location-based service providers) only serve a specific geographical region. And certainly, not all services are available everywhere. That's why LoST has the concept of a service boundary (see Section 3.4.4) – to tell clients the area where a service is offered by a given provider or PSAP. Furthermore, the "listServicesByLocation" query can be used to figure out the available services at a particular location. The LoST server then returns a list of services – without informing the client the extent of the area for which the returned service list is valid. This may lead to the situation where a client initially discovers all available services by the "listServicesByLocation" query, and then moves to a different location (while refreshing the service mappings), but without noticing the availability of other services. The following imaginary example (taken from Wolf (2010)) illustrates the problem for emergency calling:

The client is powered on, does a location determination (resulting in location A) and performs an initial "listServicesByLocation" query with location A requesting urn:services:sos. The LoST server returns the following list of services:

```
urn:service:sos.police
urn:service:sos.ambulance
urn:service:sos.fire
```

The client does the initial LoST mapping and discovers the dial strings for each service. Then the client moves, refreshing the individual service mappings when necessary as told by the service boundary for each service. However, when arriving in location B (close to a mountain), mountain rescue service (urn:service:sos.mountain) is available, which was not available in location A. The client does not detect this, because only the mapping of the initially discovered services (police, ambulance, fire) are refreshed. Consequently, the dial string for the mountain rescue is not known by the client, and the client is unable to recognize an emergency call when the user enters the dial string of the mountain rescue. This would cause a call for mountain rescue to fail altogether.

Note that a cache of service boundaries (i.e., the service regions for individual services) cannot be considered an indicator for the region in which a specific service list is valid. The service list can change, even within the service boundary of another service. For example, the ambulance mapping is valid for a whole state, but a part of the state might also have mountain rescue service. Since the LoST protocol employs the service boundary concept in order to avoid having clients continuously trying to refresh the mapping of a specific service, a service list boundary mechanism described in Wolf (2010) provides similar advantages for service lists. Even though this document describes the service list boundary extension as optional, it could be very valuable, especially in countries with multiple emergency services as described in the imaginary example above. In countries where only one emergency service is available, the problem would only arise when moving across the border to a neighboring country offering different services.

For implementations, the attempt to implement service list boundaries should be relatively low, especially for clients that are already able to evaluate service boundaries. The service list boundary is just like another boundary; the only difference is that when a mobile client moves outside this boundary, a new listServiceByLocation query has to be trigged. The modifications to the LoST protocol are also small. The service list boundary could be simply requested by a serviceList-BoundaryRequest element, either by value or by reference. Of course, this concept has also to be supported by server implementations. Calculations on the server side of the (largest possible) service list boundary can be relatively complex. However, smaller areas can also be returned, which would result in a few more queries but would still accomplish the main purpose of the service list boundary, namely making sure that a client notices a change in the service list. More implementation considerations as well as details are available in Wolf (2010). At the time of writing, the service list boundary is not yet issued as an RFC, hence it is listed in the ongoing work chapter. The advantages of the service list boundary should outweigh the implementation effort to ensure that a client does not miss a change of available services when moving.

8.2.6 Order of Location Configuration

Since there are several mechanisms for location configuration, the question arises of the order in which the different approaches should be performed, and which of the returned location values should be used for LoST and conveyed to the PSAP (in the case that multiple location values are returned). For example, it is possible to get civic location information via DHCP, geodetic location information via HELD, and more civic location information from LLDP-MED. In such a case, which location information should be used?

This question has been discussed a little within the IETF and also the Broadband Forum. Should phones try LLDP-MED, then DHCP, then HELD (as recommended by the Broadband Forum), or in any order, or all in parallel? All of these location configuration processes may be initiated without waiting for the others to complete. As soon as valid location information is returned, it can be used and the other steps may be cancelled. The document WT-164 by the Broadband Forum has further guidance on this topic and has some flowcharts that define specific procedures (*Requirements for CPE in Support of Accessing Emergency Services (work in progress)*, 2010).

8.2.7 *Notifying Users of Emergency Calls*

Should a user be notified that he has just entered an emergency dial string? It may happen that the number was dialed by mistake, or that the user was not aware of the fact that this number is actually an emergency number in the visited country. A uniform notification to the user could be helpful here. However, care has to be taken in choosing such a notification; a simple alert saying, "You have dialed an emergency number, do you really want to place an emergency call?" could cause confusion, and would cause problems for non-English speakers. In addition, a user in distress might choose the wrong option, thus preventing the emergency call from taking place.

8.2.8 *Connecting Emergency Dial Strings and Emergency Authorities*

It is entirely possible for multiple LoST mappings for different services, with different contact URIs, to be supplied with the same emergency dial strings. At the moment, it is unclear how to deal with different emergency services that are reachable via the same emergency number. Which emergency service should be selected when the user dials the number? Closely related to this situation is when there exist multiple emergency dial strings for a particular emergency service. The XML element serviceNumber must contain only one dial string, so if some jurisdictions have this problem, there may be a need to extend the LoST protocol.

It is also not clear if for every location there has to be an emergency service urn:service:sos even if there is not a single consolidated emergency number, with emergency services instead being reached by separate mappings for urn:service:sos.police, urn:service:sos.ambulance, and urn:service:sos.fire, for example.

The scenarios shown above could exist in reality. For example, even the reference implementations of LoST servers shows multiple emergency services with the same emergency dial string in the sample data. An example from Austria: The ambulance and the water rescue are both alerted by calling 144. If the water rescue is supplied in LoST as a separate emergency service, then there would be two services with the same service number. When both calls should reach the same PSAP, the distinction is not that important. However, routing calls for the ambulance service to the floodwater rescue differently would be problematic.

8.2.9 *Disconnection during an Emergency Call*

Callers in distress situations might try to end the call (on purpose or by accident) even though the PSAP call taker still has questions on the emergency situation, or the caller might hang up before a call taker is even able to answer the call. The core ECRIT specification (Rosen and Polk, 2010) requires that the user be prevented from ending an established call, reserving this capability for the PSAP. However, at the moment it is not quite clear how to implement this feature. More detailed requirements are collected in a new document (Rosen, 2009).

8.2.10 *LLDP-MED ELIN will not be Supported*

An ELIN is a valid E.164 phone number and is used in North America for the identification of emergency calls. Therefore, LLDP-MED is capable of using ELINs as part of the location configuration process. Unfortunately, the ECRIT architecture does not support the use of ELINs at all. This may lead to the unacceptable situation that a client has successfully configured location information (as an ELIN), but is unable to place an emergency call, since the IETF has not standardized the use of ELINs. ELINs can neither be conveyed in PIDF-LOs nor can they be used to figure out the responsible PSAP via LoST.

So right now we would not recommend using ELINs when wanting to access the ECRIT emergency calling architecture. The reason why the IETF did not consider ELINs is because they are part of the legacy of emergency call system and probably not part of the Internet-based future of emergency calls.

8.2.11 *Civic Boundaries*

LoST makes a lot of use of the concept of geospatial boundaries (see, e.g., Section 3.4.4). Besides geodetic descriptions, most commonly in the form of polygons, a boundary can also be given in civic form, at least according to the LoST protocol. Unfortunately, exactly how one would specify such a boundary is rather vague in the current documents, especially with regard to how to decide whether a given civic address is inside a civic boundary. The two obvious strategies – giving a single high-level entity (e.g., a state) or a list of valid civic addresses – are either not general enough or require impractical amounts of data to be stored and transmitted.

This open issue is addressed in a draft that has been submitted to ECRIT (Thomson and Wolf, 2010). The document provides a concrete definition of a civic address boundary and an algorithm for matching a given civic address against a boundary. Namely, according to the draft's definitions, a civic address boundary has the same structure as an individual civic address, and an address is contained within a boundary if every element of the address matches the corresponding element in the boundary, if one exists.

It remains to be seen whether this will be a practical scheme to implement. Civic addresses may have a complex structure, as the necessity of civic address considerations demonstrates (see Section 3.2.1). Moreover, since the algorithm is unable to take into account every civic address consideration document for each country in the world, an accurate description might not always be possible for every boundary one might want to describe. Especially when civic address fields are concatenated into a single PIDF-LO element (as might happen with some civic address considerations), mapping constraints on these fields to the strict equality required by the document could be difficult or impossible. When using the algorithm to figure out whether a civic address is within a civic boundary, false negatives may be inevitable for some addresses. However, this may have little impact and may only lead to more frequent queries to the LoST server, refreshing mappings unnecessarily.

8.2.12 LoST Service Boundary References and Location Types

A LoST service boundary specifies the area for which a mapping is valid (see the explanation in Section 3.4.4). In cases where the client requests the boundary information by reference, they will get a key back, which can be used to perform a separate query for the value. The key identifies the service boundary uniquely within the scope of the LoST server hosting it (refer to Section 5.6 of RFC 5222). The advantage of using the reference is that a client may cache the boundary information and does not have to request it each time. As long as the key does not change, the service boundary will not change either.

To enable the client to make use of the boundary information, the boundary needs to be in a format that the client is able to evaluate; if the client has geodetic location, it might not be able to evaluate a service boundary in civic format, and vice versa. When a client requests a service boundary by value in a "findService" query, the server knows which type of location information the client has included in the request in order to use that type to include the boundary information in the response. However, a "getServiceBoundary" query is decoupled from the initial findService request and does not contain any hint as to which location information the client is able to use. There is no way to tell the server how the client would like the boundary information.

The only way to do solve this issue right now is to have the hint about the format of the location information in the key that the client provides to the server when it acquires a service boundary by value with a "getServiceBoundary" query. The key would have to be used not only to identify a certain boundary but also the type of location information. So the server actually has to maintain three keys for each logical boundary: One key for civic, one for geodetic-2d, and one for clients that understand both. (RFC 5222 says that the boundary information should be returned in all profiles the client is able to understand.) And for every other profile that is registered with IANA, the number of keys necessary will grow combinatorially – every possible combination of types has to be supported.

Currently, the location profile for the service boundary has to be implicitly in the key. If a client does not get back the appropriate profile, it cannot evaluate the service boundary and all the optimization that the concept of the service boundary offers would be undone.

8.2.13 Emergency Calls to Counseling Services

Are counseling services emergency services or not? In some jurisdictions they are, in others not. Sometimes only the services under urn:service:sos are considered as emergency services but actually there are also emergency services under urn:service:counseling. So it is crucial not to forget to also to perform LoST mappings for these counseling services.

For example, the ECRIT framework document only mentions urn:service:sos when talking about how to mark emergency calls (e.g., Section 3). So one might get the incorrect impression that calls to counseling services are less important, even though they are considered an emergency call in several jurisdictions and may have unique dial strings assigned. (E.g., in Austria there are two counseling services, reachable with the emergency dial string 141 and 147; see Table 5.2.) These calls would also have to be detected and correctly routed by VoIP-based emergency calling devices. It's worth keeping in mind that calls to counseling services may help to save lives, just like calls to the police or the fire brigade.

8.3 Ongoing Implementation Issues

There are a few minor issues that have caused problems with some initial implementations of ECRIT emergency calling. Developers of new VoIP equipment should keep these issues in mind.

8.3.1 Service URNs as Request URIs

The ECRIT architecture makes extensive use of the recently standardized Service URNs for emergency calls. However, when trying to use a Service URN in a SIP message as a request URI, one can easily get into trouble with many current SIP implementations. Most software implementations are not able to handle Service URNs at all and might reject such a request. Hence, an emergency call could fail, even though the client has done everything properly, simply because the call is marked as an emergency call (by the Service URN) and is routed via a proxy server without support for emergency calls. When the call fails via a proxy, the client could try to establish the connection directly to the PSAP.

8.3.2 Converting from the DHCP Location Format to PIDF-LO

The DHCP options do not carry location information as a PIDF-LO. Consequently, a conversion is necessary before the location can be sent in a SIP message. The current definition for geodetic location (RFC 3825 (Polk et al., 2004)) leaves a lot of ambiguity in how this conversion should be performed. (Civic addresses elements in DHCP use the same name as the XML elements, so the conversion is straightforward.) The only way to specify geodetic location information in DHCP is with latitude, longitude, and altitude, along with the resolution values for these parameters. However, in PIDF-LO there are several shapes (see Section 3.2.1), like polygons, ellipses, and prisms. In particular, the GML format used for geodetic information in PIDF-LO has no notion of resolution. The GEOPRIV working group in the IETF has created a document to clarify this ambiguity (Polk et al., 2010), which should soon be published as an RFC.

8.3.3 LLDP-MED Difficulties

LLDP is a layer 2 protocol and uses LLDP multicast addresses. So most operating systems require a raw socket and therefore administrative permissions in order to be able to send and receive LLDP. If a SIP client wants to use LLDP-MED for location configuration, it would have to run with administrative permissions as well.

So it could be advantageous to have direct support for LLDP-MED location configuration in the operating system. In the long term, it would be helpful for the whole location configuration process (with multiple protocols) to evolve into an operating system service, so that other application programs could use location information.

Another note regarding LLDP: not all currently deployed network devices handle LLDP multicast addresses correctly (e.g. some switches drop them, others send them out unmodified on all their ports). Networks that want to use LLDP would need switches with correct LLDP-MED support.

8.3.4 Multi-Part SIP Bodies and Message Size

Multi-part SIP messages are often needed in emergency calls in order to accommodate the session description (needed to establish the session) and location information (by value) at the same time. Unfortunately, not all SIP implementations can handle these messages – some actually have hard limits on how big a SIP message can be.

Furthermore, the creation of such multi-part SIP messages increases the message size significantly. SIP is often carried in UDP, but RFC 3261 requires the use of TCP when a message's size exceeds 1300 bytes. As message sizes grow, this requirement might be problematic, since not all SIP nodes allow TCP connections (although an increasing percentage do). In cases where TCP support is needed but not present, users might not be able to establish a connection, which would be unacceptable if the large message in question is establishing an emergency call. Moreover, a check on the SIP message size is often part of security checks, with unusually large messages considered suspicious and dropped. Under this criterion, a SIP message containing several PIDF-LO objects might easily be considered suspicious. So these issues have to be considered when operating a SIP proxy. (Actually, all these issues can be avoided if the call setup is done directly with the PSAP, rather than via a proxy server, although this might not be practical in all cases.)

References

Polk J, Schnizlein J and Linsner M (2004) Dynamic Host Configuration Protocol Option for Coordinate-based Location Configuration Information. RFC 3825.

Polk J, Schnizlein J, Linsner M, Thomson M and Aboba B (2010) Dynamic Host Configuration Protocol Options for Coordinate-based Location Configuration Information. Internet Draft (work in progress).

Requirements for CPE in Support of Accessing Emergency Services (work in progress) (2010). Technical report, Broadband Forum.

Rosen B (2009) Requirements for Handling Abandoned Calls and Premature Disconnects in Emergency Calls on the Internet. Internet Draft (work in progress).

Rosen B and Polk J (2010) Best Current Practice for Communications Services in Support of Emergency Calling. Internet Draft (work in progress).

Schulzrinne H, McCann S, Bajko G, Tschofenig H and Kroeselberg D (2010) Extensions to the Emergency Services Architecture for Dealing with Unauthenticated and Unauthorized Devices. Internet Draft (work in progress).

Thomson M and Wolf K (2010) Describing Boundaries for Civic Addresses. Internet Draft (work in progress).

Wolf K (2010) LoST Service List Boundary Extension. Internet Draft (work in progress).

9

Summary and the Outlook for the Future

Emergency calls are surely one of the most critical functions of any telephone system. People take for granted that they can pick up a telephone and place an emergency call, and the simple existence of this service helps them feel more secure. Even if most VoIP customers aren't explicitly asking their VoIP providers to support emergency calling, if they find themselves in an emergency, they will naturally reach for their "phone" – especially if the end device looks like a traditional telephone. Which technology a phone uses makes no difference to the user, and sometimes the difference between a normal phone and a VoIP phone isn't even apparent at first glance (e.g., with an analog telephone using a VoIP adapter).

Unfortunately, right now emergency calls are not always handled optimally. In addition to the challenges that VoIP introduces, there are also opportunities to improve how traditional, circuit-switched calls are handled. In order to serve VoIP calls with equal or even better quality to circuit-switched calls, the Internet standards community has developed the ECRIT architecture discussed in this book. Since the proportion of calls and subscribers that use VoIP is continuously increasing, there is an urgent need to implement this architecture. The standards for the ECRIT architecture are essentially complete, with just a few details to be worked out, and national standards bodies are already developing architectures that apply the ECRIT architecture to their jurisdictions. Organizations that are involved in the emergency calling process – including PSAPs, telecommunications companies, ISPs, and others – should be considering how they will support VoIP emergency calling, and become familiar with the ECRIT architecture as well as implementation approaches. Finally, the overall task of improving

VoIP Emergency Calling: Foundations and Practice Karl Heinz Wolf and Richard Barnes

emergency calling capabilities will require the cooperation of several different organizations (as outlined in Section 5.1).

The most important considerations for an emergency calling system are:

- availability of local emergency numbers;
- connection to the PSAP that is responsible for the caller's current location;
- reachability of emergency numbers free of charge, including reachability for callers with late payments or no credit;
- identification and geolocation information about callers to enable efficient emergency response.

Location information is an essential component of an emergency call. Because of the increasing mobility of users, as well as the separation between the IP access providers and VoIP service providers, determining the location of a VoIP caller presents both technical and organizational challenges. The different possibilities for location determination and location configuration are laid out in Chapter 4.

Two steps in the emergency calling process require location information:

1. Determining the responsible PSAP (as well as local emergency numbers).
2. Response planning and dispatch (enabled by automatic delivery of location to the PSAP).

Without location information, there can be no emergency calls, since in order for the caller's phone to even recognize that an emergency call is being placed, it needs to know the local emergency numbers (e.g., 133 in Austria or 911 in the US), and it needs to know the location to look them up. Automatically delivering location information to PSAPs will be a critical improvement, for circuit-switched calls as well as VoIP calls.

An overall picture of the ECRIT architecture, including all the relevant protocols, standards, and Internet drafts (the "big picture") can be found in Chapter 3. In order for future emergency calling systems to have the same level of global interoperability that the Internet enjoys today, it is critical that the IETF standards for the ECRIT architecture be followed and faithfully implemented.

It can be expected that the other types of emergency communication may also benefit by adopting modern communication channels. Take early warning, the emergency communication from an authority to the individuals as an example. Today's early warning systems, based on cell broadcast or even simple sirens, could be supported by Internet-based methods as well. Work on this topic has already started in the IETF, to be conducted within the ATOCA working group. This group will produce an emergency alerting architecture that fits well into the ECRIT emergency calling architecture.

ECRIT emergency calling implementation is not difficult or far in the future, though. In Chapter 6, we discussed several prototype software implementations of different parts of the architecture, as well as some basic exercises that operators can go through to get familiar with it. There are even some practical steps toward making production networks ECRIT-compliant, with little or no new investment.

Especially given the interactions that will be necessary between emergency service operators and access network operators (i.e., ISPs), there will likely be a need for local coordination efforts in many regions. Performing interoperability tests of prototype systems, even while standards are still being finalized, can provide valuable experience. Looking a little more into the future, it is possible that the need for coordination will create a need for new telecommunications regulations – especially if the telecom industry doesn't come together on its own to put forward solutions for national requirements. (Such industry efforts are already in progress, e.g., in the US and Switzerland.)

Implementing the ECRIT architecture will certainly require some investment by network operators in new infrastructure and training, particularly in developing location services to provide information about subscribers. However, the uses of location information go well beyond emergency services, and commercial applications for location have the potential to create new revenue streams for ISPs.

The ECRIT architecture developed by the IETF has the potential to fundamentally improve the state of emergency calling systems, for users of the PSTN as well as VoIP subscribers. Location information will be delivered directly to the PSAP, where it is urgently needed, and the fundamental mobility of VoIP subscribers will no longer keep them from accessing emergency services. As the number of VoIP subscribers grows, the need for VoIP emergency calling grows even more urgent.

Index

Printed and bound by CPI Group (UK) Ltd, Croydon, CR0 4YY

16/09/2026

14963383-0001